AF334691

English Language Edition

Translated by Irwin Schwartz
in collaboration with
Accurapid Translation Services, Inc. and The Language Service, Inc.
Poughkeepsie, New York

Typography by AB Typesetting
Poughkeepsie, New York

Library of Congress Cataloging-in-Publication Data

Jun-Ichi, 1936–
ō *kyōyaku kōgaku.* English]
ase conjugate optics/by Jun-Ichi Sakai.
p. cm. — Advanced science and technology series
anslation of: *Isō kyōyaku kōgaku.*
cludes bibliographical references and indexes.
BN 0-07-054315-1
Optical phase conjugation. I. Title. II. Series
16.3.067S2513   1992
2—dc20

92-7103

ADVANCED SCIENCE AND TECHNOLOGY

PHASE CONJUGATE OP

by Jun-Ichi Sakai

Sakai
[Is
Ph

Ti
In
I9
1.
QC4
535'.

McGı
New York   St. Louis   San Francisco
Caracas   Lisbon   London   Madri
Montreal   New Delhi   Paris   Sa
Singapore   Sydney

Preface

Optics has a long history. The laws of refraction, Fermat's Principle, and spectrum analysis were already known by the 17th century. The disciplines of geometric optics and wave optics fall under the category of "classical optics," for which Born and Wolf's *Principles of Optics* has become the bible. Maxwell elucidated that light was a type of electromagnetic wave, and that it could be described in terms of its amplitude and phase. In addition, the management of spatial information is often cited as a feature of light. However, since phase data become degraded as a function of the characteristics of the recording media, only optical intensity data retain their integrity when spatial information is recorded. Therefore, the problem of phase reconstruction persists, in which the original phase information must be searched for among the optical intensity data. Historically, the concept of phase in optics has been used as first-order data, so to speak, as in the description of polarization, but not as spatial information.

Since Maiman invented laser optics in 1960, research into making the best use of the characteristics of the laser — coherence, high luminosity, and directionality — has broadened. Spectral diffraction and optical nonlinearity are areas of investigation. New nonlinear phenomena have been observed as a result of the high luminous intensity of lasers. Among these phenomena are frequency upconversion, parametric amplification and oscillation, stimulated optical scattering, optical echoes, two-photon absorption, and self-focusing. In 1972, Zel'dovich et al. discovered optical beam correction applications through experiments with stimulated Brillouin scattering. These phenomena have been explained by the introduction of the concept of phase conjugation into the optical arena, corresponding to complex conjugates in mathematics. That is, it was discovered for the first time that another wave with inverted phase information can be generated when a wave with certain complex amplification exists. The new phenomenon is called phase-conjugate light or the phase-conjugate wave. Based on the appearance of this phase-conjugate light, a new branch of optics called "phase-conjugate optics" has materialized. Now the field of optics can deal with spatial phase information in addition to light intensity. "Phase-conjugate optics" was established in name and in substance after the degenerate four-wave mixing technique was proposed by Hellwarth in 1977; he was followed by many others who established the experimental procedures in this field.

The objective of this book is to organize and introduce the fundamental issues related to the basic concepts of phase-conjugate light and its applications. In Chapter 1 we explain the history of phase-conjugate optics. Chapter 2 describes the basic concepts and characteristics of phase-conjugate light. Chapter 3 unfolds with Maxwell's equations as the starting point for the theory of phase-conjugate light generation. Classical methods based on applied coupled-mode equations are introduced and developed. Optical Kerr media, along with photorefractive media, are considered here as media for generating phase-conjugate light. Four-wave mixing is explained as having become the now-standard means for generating phase-conjugate light. Quasi-classical and quantum-theoretical treatments are touched upon in Chapter 4, and their relationship to classical theory is clarified. Chapter 5 introduces the theory of non-four-wave mixing. Chapter 6 is divided into types of phase-conjugate light-generating experiments; common elements and methodologies within each type of experimental technique, and issues dealing with peculiarities of media are discussed. Chapter 7 explains optically nonlinear media from the perspective of phase-conjugate generating media. Chapter 8 shows actual examples of the special characteristics of phase-conjugate light, and discusses areas of and formats for application. Finally, the future directions of phase-conjugate optics are briefly touched upon.

The features of this book are as follows: (i) The nonlinear refractive indexes for nonlinear media in the theory of phase-conjugate light generation are all written in SI units. (ii) Notation is made uniform throughout chapters and sections. (iii) We reflect an awareness of the changes in the fundamental theoretical equations in the theory of phase-conjugate light generation due to its variety of sub-fields (the effects of absorption and detuning, quasi-classical theory, quantum theory). (iv) Since each chapter is made as independent from the others as possible, each can be read and understood by itself. (v) Formulas for conversion between SI and cgs esu units for nonlinear refractive indexes are given in the appendix.

There are a particularly large number of equations in this book. Although we have paid special attention to the derivation of the equations and the uniformity of symbols, there is always the risk that not only trivial typographical errors may remain, but that there are some serious errors.

The author was on loan from NTT as an associate professor at Tokyo University's Advanced Technology Research Center from 1988 to 1989. This book was written during my stay at the university under Professor Ogashi's encouragement, and extended throughout my research tenure. Utmost gratitude goes to the NTT executives who gave me this precious opportunity, to the professors — particularly Professor Ogashi and Assistant Professor Yoshio Hodate — and to the staff of the Research Center. I am deeply grateful to the editing department at Asakura Publishing Company for helping me complete this book.

October, 1990

Jun-Ichi Sakai

How to Use This Book Effectively

Readers of this book are those who want to quickly learn about phase-conjugate optics. Accordingly, a guide for reading this book with that type of reader in mind is presented below.

At a minimum, Chapter 2 and Sections 3.1 and 3.2 should be read for an overall understanding of phase-conjugate light.

In addition, readers who are interested in phase-conjugate light experiments should understand the contents of Chapter 6. In particular, Section 6.2 presents the basic technology for generating phase-conjugate light, which can be used as a point of reference by those who are planning to conduct experiments. Information on phase-conjugate light generating media is covered in Chapter 7.

Readers who have an interest in the application of phase-conjugate light will want to look at Chapter 8 since it explains examples of the characteristics of phase-conjugate light. In addition, Chapter 6 should be read and understood.

Readers interested in the theory of phase-conjugate light generation should read through Chapters 3–5. They should be able to get a good detailed feel for aspects of the elaboration of the theory. Since Section 4.2 is specialized, it may be ignored with no impediment to understanding.

Preparation for Using This Book

1. Units

Although both the cgs and SI (Système International) unit systems are used for electromagnetism, the SI system is basically used in this book. The SI units for all symbols are clearly stated in Sections 3.1, 3.2, and 3.3. Since cgs esu units are frequently used in the discussions relating to experiments, unit values from both systems are written side by side in these cases. However, in cases where conversion is difficult — as for the linear index of diffraction — values are given in cgs and esu units only. The relationships among wave equations using both unit systems are shown in Appendix A, and conversion factors are shown in Table A.

2. Symbols in Equations, Illustrations and Tables

Equations and tables are double-numbered. The first numeral represents the chapter; the second numeral the positional ordering within the chapter.

3. Reference Bibliography

The bibliographical references are numbered for each chapter, and are summarized at the back of the book.

4. Terminology and Symbols

Phase-conjugate optics has a short scientific history and, despite some non-uniformity of terminology, there is generally fixed usage. Only the refractive index and the photorefractive effect are written in parallel. We have tried as best as we could to make symbols that express the same thing as uniform as possible within chapters and sections. Because some symbols are used frequently, the same symbol may have various meanings. The symbol n generally denotes the refractive index. In Section 3.3, the carrier density is n_c, and is distinguished from the refractive index. Following are explanations of other notations.

Notation

$\hat{a}_j$	Boson operator
A_j	Electric field complex amplitude ($j = f$, forward pump wave; $j = b$, reverse pump wave; $j = p$, probe wave; $j = c$, phase-conjugate wave)
B	Magnetic flux density
c	Speed of light in vacuo
D	Electric flux density
e	Quantity of electric charge
E	Electric field
E_{st}	Static electric field
h	Planck's constant
H	Magnetic field
i	Imaginary unit
I	Light intensity
I_{sat}	Saturation light intensity depending on detuning
I_{sat}^{0}	Saturation light intensity at the absorption nucleus
J	Current density
k	Wave vector
k_B	Boltzmann constant
k_G	Diffraction grating vector
K	Coupling coefficient
L	Medium thickness or length
m	Light-intensity spatial modulation index
n	Refractive index
n_2	Optical nonlinear refractive index
n_c	Carrier density
N_A	Acceptor density
N_D	Donor density
N_D^{+}	Ionized donor density

P	Polarization
P_{NL}	Nonlinear polarization
$\boldsymbol{r}$	Position vector
r	Ratio of reverse pump-wave intensity I_b to forward pump-wave intensity I_f (I_b / I_f)
R	Phase-conjugate light coefficient of reflection, wavefront curvature radius
$\vec{R}$	Rank 3 electro-optic tensor
T	Probe wave penetration, absolute temperature
α	Light intensity absorption coefficient, coherent state (Section 4.2)
α_0	Small-signal light-intensity absorption coefficient
δ	Standardized detuned frequency at incident light frequency ω, atomic resonance frequency ω_R, and horizontal relaxation time T_2 $[=(\omega - \omega_R)T_2]$
δk	Phase mismatch
ε	Electric permittivity of a medium (at optical frequencies)
ε_0	Electric permittivity of free space
ε_s	Static electric permittivity
$\varkappa$	Coupling coefficient
λ	Wavelength in vacuo
μ	Carrier mobility
μ_0	Magnetic permeability of free space
ρ	Electric charge density, density matrix (Section 4.2), medium density (Section 5.2)
σ	Electric conductivity
$\chi^{(n)}$	Coefficient of nth order optical nonlinearity
ψ	Phase difference between optical intensity distribution and spatial electric field distribution
ω	Optical angular frequency
ω_R	Atomic resonance frequency
Ω	Probe detuning frequency corresponding to pump wave detuning frequency

Table of Contents

1.

The History of Phase-Conjugate Optics

1.1 Methods of Generating Phase-Conjugate Light

The concept of "phase-conjugate light" was first precisely described by the Russian Zel'dovich and his colleagues.*[1] In 1972, through stimulated-Brillouin backscatter experiments using methane gas, they discovered the experimental result that narrow-beam spots with little diffraction spread could be obtained if light was passed through the same distorting (aberrating) medium twice. The result of analyzing the clear spot obtained was that phase correction automatically occurred due to the nature of phase-conjugate light. Although the facts surrounding phase correction were discovered in stimulated Raman scattering also, the flaw with such phase correction was that it was imperfect since these scattering phenomena were accompanied by a frequency shift.

Subsequently, in 1976, three-wave mixing was proposed as a phase-conjugate light-generating technique that was not accompanied by a frequency shift.[3]

---

* The existence of conjugate-light phase correction predates this. In 1971, through the discovery of dynamic holographic processing by Stepanov et al., holograms were simultaneously recorded and reconstructed.[2] A Q-switched ruby laser as light source and a solution of GaCl and pthalocyanimine dissolved in chlorobenzene were used in what is now called the direct-response degenerate four-wave mixing format (though its advantages described below were not recognized). There was no awareness of the concept of phase-conjugate light.

This, it was argued, was a means for recovering degraded image quality when optical images were conveyed by multi-mode optical fibers. Few experimental results are available because of the required light and source conditions, phase-matching conditions, and the limitations of nonlinear optical media, as discussed in Section 5.1.

Degenerate four-wave mixing was proposed by Hellwarth in 1977 as a technique for eliminating the flaws described above.[4] Here, two pump waves injected light of the same frequency ω from opposite directions through a medium having a third-order optical nonlinearity. Then, when a third wave (probe wave) of frequency ω was injected, phase-conjugate light of the same frequency emerged in the opposite direction. The term "degenerate" is derived from the four light waves all having the same frequency. Since degenerate four-wave mixing has a variety of advantages — this is discussed in Section 3.1 — many experiments can be performed using gases, liquids, dielectrics, semiconductors, organic materials, and non-crystalloids, making it the standard experimental method for generating phase-conjugate light. Degenerate four-wave mixing was proposed at almost the same time by Yariv et al.[5] Here, the analysis of phase-conjugate light through coupled-mode theory became the standard technique.

The experiment that first suggested the practical value of degenerate four-wave mixing was performed in 1977, and used CS_2 as the optical Kerr medium.[6] With a Q-switched frequency-doubled Nd:YAG laser as light source, phase-conjugate light was employed in a test-chart image reconstruction experiment. The result was a reconstructed image with a spatial resolution of 40 lines/mm. Other optical Kerr media, such as Na vapor and semiconductors, were also employed.

The photorefractive effect is often used for the generation of phase-conjugate light through four-wave mixing. The photorefractive effect was discovered in 1966 during research into satisfactory holographic optical memory media when a low-power beam (a few mW) was used to illuminate $LiNbO_3$ and $LiTaO_3$ and produced abnormal refractive-index changes,[7] although this phenomenon was considered to be bothersome at the time. In addition to the crystals mentioned above, the phenomenon is well understood for $BaTiO_3$, $Bi_{12}SiO_{20}$ (BSO) and for $Sr_{1-x}Ba_xNb_2O_6$ (SBN), as well as for other compounds. Incidentally, the term "photorefractive" was derived by analogy with "photochromism."

The first photorefractive medium used for the generation of phase-conjugate light was BSO, in 1979; an argon-ion laser was used as the light source.[8] Subsequently, phase-conjugate light was also generated with $BaTiO_3$. Phase-conjugate light generation via a self-excitation arrangement not using independent pump beams was realized with $BaTiO_3$ in 1982.[9] $BaTiO_3$ has a big advantage because of its high electro-optic coefficient which causes the scattering of light within crystals as well as auto-oscillation between external resonators to create a pump wave. An He-Ne laser (632.8 nm) was used as the light source in this experiment, and at least $50\mu W$ of phase-conjugate light power was generated.

Resonance absorption, two-photon absorption, optical echoes and thermal effects are also used in the generation of phase-conjugate light.

1.2 Applications of Phase-Conjugate Optics

Although phase-conjugate light has a variety of characteristics, the application that received the most immediate attention was phase correction. A beam transmitted in air is perturbed by atmospheric fluctuations on the path to the target. Initially, adaptive optics were used to achieve phase correction. A reference beam was sent from the target to the transmitter and the transmission path distortion was measured by many subapertures located at the transmitter. Before the light beam was emitted, a deformable mirror reversed the distortion along the transmission path and offset the distortion following transmission. It is important for the mirrors to be distorted accurately by this technique, prompting the use of various types of sensors and servomechanisms for the distortion to be precisely measured. In addition, spatial resolution was restricted by the number of subapertures and phase shifters that could be set up. In this way, distortion measurement and reverse distortion could be performed simultaneously and with high precision.[10] This type of phase correction used in optical resonators, laser fusion, and optical lithography, among other applications, is an area where phase-conjugate optics might replace adaptive optics.

One aspect of degenerate four-wave mixing is directed toward real-time holography. As our knowledge becomes refined, we are learning that many similarities exist between them.[11] Consequently, the same emerging ideas that are being used as the basis for conventional holography are also being used for phase-conjugate optics. They are used for optical data storage in computers and associated memory.

There is also basic research being performed in nonlinear spectra, optical pulse compression, and the generation of optically-squeezed modes using the interaction between phase-conjugate light and media.

Phase-conjugate light has a deep relationship with holography, optical echoes, optical bistability, chaos, and other phenomena. In addition, there are dramatic developments in nonlinear organic materials and quantum-well structures as substances having large coefficients of nonlinear optical susceptibility. Phase-conjugate optics is expected to progress in a way that will complement the growth in insight and technology in other areas. As a result, phase-conjugate optics will progress by making use of the unusual characteristics of phase-conjugate light while it is simultaneously developing as one branch of nonlinear optics as well as an independent disciplinary area.

2.

Basic Concepts and Properties of Phase-Conjugate Light

2.1 Basic Concepts of Phase-Conjugate Light

For simplicity, consider a monochromatic plane wave (probe wave) of frequency ω whose direction of propagation is $+z$. This is represented by the equation

$$E_p(r, t) = (1/2) A_p(x, y) \exp[i(\omega t - k_p z)] + \text{c.c.}$$
$$= (1/2) \Phi_p(r) \exp(i\omega t) + \text{c.c.} \tag{2.1}$$

where $A_p(x, y)$ is the amplitude, which varies with position more slowly than the phase (slowly varying amplitude); k_p is the wave number of the incident wave in the z direction; r is the position; and c.c. is the complex conjugate of the preceding expression. $\Phi_p(r)$ contains polarization data as well as spatial data. The reflected wave is described by

$$E_c(r, t) = (1/2) A_c(x, y) \exp[i(\omega t - k_c z)] + \text{c.c.}$$
$$= (1/2) \Phi_c(r) \exp(i\omega t) + \text{c.c.} \tag{2.2}$$

where k_c is the reflected-light wave number in the z direction. If $\Phi_c(r) = \Phi_p^*(r)$, that is, if $A_c = A_p^*$ and $k_c = -k_p$, the reflected wave E_c is called the phase-conjugate wave of incident wave E_p.[1,2] As discussed below, phase-conjugate waves are not completely limited to light waves, but since this is the domain of this book, we will use the terminology "phase-conjugate light." Waves (2.1) and (2.2) considered as functions of position are complex conjugates. Furthermore,

$$E_c(\boldsymbol{r}, t) = E_p(\boldsymbol{r}, -t) \tag{2.3}$$

This indicates that the shape of the wavefront here remains the same and that there is inversion only with respect to the time axis. The phase-conjugate light here is said to be a time-reversed wave. Since the two waves described by the above phase conjugation relationships behave as if they were reflected by a conventional mirror (but with a law of reflection different from that for a normal mirror, as explained below), the incident and reflected light can be thought of as being related via a "black box" called a phase-conjugate mirror.

2.2 Properties of Phase-Conjugate Light

Figure 2.1 illustrates how an undistorted phase-conjugate light image is produced when a plane wave traverses a phase-distorting medium. Initially plane wave 1 passes through glass having a higher refractive index at the center than at the edge. The result is a delay 2. If this wave were incident on a conventional mirror, the reflected wave plane 3 would become reversed from 2 since the direction of propagation is reversed at the surface of the mirror as a result of the folding back of the wave. If the light passes through the same distorting medium a second time, the phase delay at the center doubles. On the other hand, if the wavefront 2 is incident on a phase-conjugate mirror, the reflected wave 3 has the same shape as the incident wavefront and only the direction of propagation is reversed, as was seen in Equation (2.3). If this wave again passes through the same distorting medium, there is a phase delay at the center and the phase delays coming and going through the glass offset each other. Because of this, spatial phase correction is a feature of phase-conjugate light. This property has a variety of applications, as discussed below.

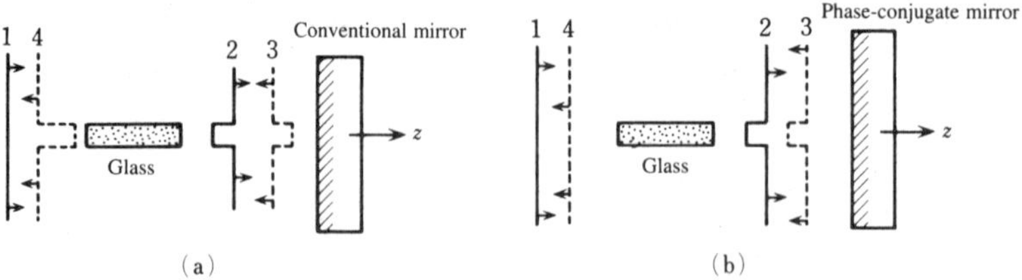

Figure 2.1 Phase correction by phase-conjugate light illustrated for the case of a plane wave.

(a) Reflection by conventional mirror, (b) reflection by phase-conjugate mirror. 1: Wavefront before incidence on glass, 2: Wavefront after passing through glass, 3: Wavefront after reflection by mirror, 4: Wavefront after passing through glass a second time.

Next, we give a full analysis of the electric field of phase-conjugate light. Its properties will be explained in some detail. A phase-conjugate mirror is described by the following complex reflectance

$$R \exp(i\phi_{pc}) \qquad (2.4)$$

where both R and ϕ_{pc} are real numbers. Now for a wave propagating in the z direction, the positional dependence of the electric field is described as follows in terms of the real amplitude and phase

$$\Phi(\boldsymbol{r}) = \hat{e}A_R(x, y) \exp[i\phi(x, y)] \exp(-ik_z z) \qquad (2.5)$$

Here $\hat{e}$ is the polarization vector and k_z is the wave number in the z direction. The following results are obtained for an ideal phase-conjugate mirror.[3]

(i) $\hat{e} \rightarrow \hat{e}^*$

(ii) $A_R \rightarrow RA_R$

(iii) $\phi \rightarrow -\phi + \phi_{pc}$

(iv) $k_z \rightarrow -k_z$

Statement (i) is characteristic of a phase-conjugate mirror. For example, when clockwise (counter-clockwise) circularly polarized light is incident, the reflected wave has clockwise (counter-clockwise) circular polarization. This compares with the polarization change for a conventional mirror. In regard to the amplitude in (ii), R is always less than 1 for a conventional mirror, and the wave is attenuated. The novel feature of the phase-conjugate mirror, as discussed below, is that $R > 1$, making it possible to amplify an incident wave. Property (iii) expresses a phase inversion within a plane parallel to the mirror surface. This changes to $\phi \rightarrow \phi + \phi_M$ (ϕ_M: change in phase specific to the mirror) for a conventional mirror. Statement (iv) means that, among the

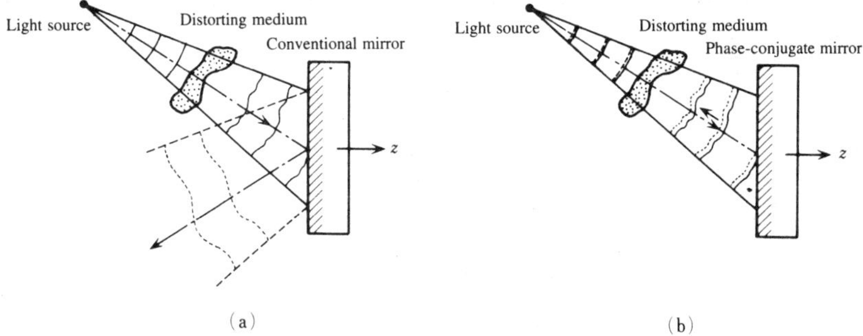

Figure 2.2 Spherical wavefront obliquely incident on a mirror.

(a) Reflection by conventional mirror, (b) reflection by phase-conjugate mirror. Solid line indicates wavefront before incidence; broken line indicates wavefront after reflection.

components of a propagating wave, the component normal to the mirror is reversed, as with a conventional mirror.

We will use an illustration to describe an example of the properties of phase-conjugate light. Figure 2.2 shows a surface wave obliquely incident on a mirror. In the case of a conventional mirror (a), the mirror simply folds back the path of the light on the surface of the mirror, and the phase distortion is propagated as is. However, in the case of a phase-conjugate mirror (b), if $\phi_{pc} = 0$, the original light path is strictly retraced after reflection from the phase-conjugate mirror even though there is phase distortion due to properties (iii) and (iv).

The following shows that phase-conjugate light satisfies Maxwell's equations. The wave equation is

$$\nabla E^2 + \omega^2 \varepsilon \varepsilon_0 \mu_0 E = 0 \tag{2.6}$$

where E is the electric field, ε is the electric permittivity of the medium, ε_0 is the electric permittivity and μ_0 is the magnetic permeability of free space, and $c = 1/(\varepsilon_0 \mu_0)^{1/2}$ is the speed of light in vacuo. If Equation (2.1) for the probe wave is substituted in (2.6), the following results.

$$\nabla^2 A_p + (\omega^2 \varepsilon \varepsilon_0 \mu_0 - k_p^2) A_p - 2ik_p(\partial A_p/\partial z) = 0 \tag{2.7}$$

If Equation (2.2) for phase-conjugate light is substituted in (2.6), then

$$\nabla^2 A_c + (\omega^2 \varepsilon \varepsilon_0 \mu_0 - k_c^2) A_c - 2ik_c(\partial A_c/\partial z)$$
$$= \nabla^2 A_p{}^* + (\omega^2 \varepsilon \varepsilon_0 \mu_0 - k_p^2) A_p{}^* + 2ik_p(\partial A_p{}^*/\partial z) = 0 \tag{2.8}$$

In comparing Equation (2.7) with (2.8), we see that they are complex conjugates. Consequently, since phase-conjugate light satisfies the wave equation, the concept of phase-conjugate waves not only applies to light waves but also to electrical waves (millimeter waves, microwaves, etc.), sound waves, and other phenomena described by wave equations.

3.

Theoretical Foundations of the Production of Phase-Conjugate Light

In this chapter we explain the fundamental theory of the production of phase-conjugate light. In preparation for this, Section 3.1 covers in broad perspective the relationships between phase-conjugate light and polarization, as well as the wave equations for polarization. Section 3.2 deals with four-wave mixing based on optical Kerr effects, but first covers techniques based on coupled-mode equations. These techniques are the basis for all other cases. After an understanding of the influences of absorption and pump-wave depletion is gained, we describe degenerate four-wave mixing on the basis of the diffraction integrals. Section 3.3 explains degenerate four-wave mixing in relation to the photorefractive effect, and gives an understanding of the band-transport and hopping models, as well as the relationship between them. The similarities of and differences between four-wave mixing and holography are shown in Section 3.4.

3.1 Introduction

(1) Generation and Polarization of Phase-Conjugate Light

Phase-conjugate light is generated by polarization resulting from the interaction between light and matter. When the incident light intensity is weak and when there are few changes in the carrier number in the substance due to the incident light, the optical polarization P [C/m^2] can be described by a power series in E [V/m].[1]

$$P = \varepsilon_0 [\chi^{(1)} E + 2\chi^{(2)} E^2 + 4\chi^{(3)} E^3 + \cdots + 2^{n-1}\chi^{(n)} E^n + \cdots] \tag{3.1}$$

In the above equation, ε_0 [F/m = C/(Vm)] is the electric permittivity of free space, and $\chi^{(n)}$ is the nth order nonlinear coefficient of susceptibility, generally expressed as a tensor. The coefficient 2^{n-1} adjusts $\chi^{(n)}$ so that $\chi^{(n)}$ can be defined as the ratio of the product of the electric field's complex amplitude and the complex amplitude of the polarization. The first-order term, proportional to E, is the Pockels effect. The second-order and higher-order terms are called nonlinear polarization terms, and are expressed by P_{NL}. Nonlinear polarization also involves frequency, and we next take a look at a few details.

If several waves having unequal frequencies are superposed, the electric field components E_m are described by

$$E_m(\boldsymbol{r}, t) = (1/2) \sum_j A_{jm}(\omega_j, \boldsymbol{r}) \exp[i(\omega_j t - \boldsymbol{k}_j \cdot \boldsymbol{r})] + \text{c.c.}$$

$$\text{(Subscripts } j \text{ distinguish types } m = x, y, \text{ or } z) \tag{3.2}$$

The sum runs over all waves j. Here, $A_{jm}(\omega_j, \boldsymbol{r})$ [V/m] is the slowly-varying complex amplitude, ω_j [s^{-1}] is the frequency, $\boldsymbol{k}_j$ [m^{-1}] is the wave number vector, r[m] is the position, and c.c. is the complex conjugate. The nonlinear polarization in this case is expressed as

$$P_{NLm}(\boldsymbol{r}, t) = (1/2) \varepsilon_0 \sum_q Q_m(\omega_q, \boldsymbol{r}) \exp[i(\omega_q t)] + \text{c.c.} \tag{3.3}$$

where $Q_m(\omega_q, \boldsymbol{r})$ [C/m^2] is the complex polarization. The second-order nonlinear polarization can be expressed by the following when the electric field and the polarization are as above.

$$Q_m(\omega_q, \boldsymbol{r}) = \sum D_2\, \chi^{(2)}_{mno}(\omega_q\,;\, \omega_j, \omega_k) A_{jn}(\omega_j, \boldsymbol{r}) A_{ko}(\omega_k, \boldsymbol{r})$$

$$\cdot \exp[-i(\boldsymbol{k}_j + \boldsymbol{k}_k)\cdot \boldsymbol{r}] \quad (m, n, o = x, y \text{ or } z) \tag{3.4}$$

A wave with frequency $\omega_q = \omega_j + \omega_k$ is generated from waves with frequencies ω_j and ω_k. When the complex conjugate of the electric field amplitude is taken, the signs of the frequency and the wave number are reversed. The degeneracy factor D_2 takes care of the case where two waves with the same frequency are present.[2] D_2 takes the value 1 for two waves of equal frequency (e.g., second harmonic) and 2 for two waves differing in frequency. $\chi^{(2)}$ [m/V] is generally a rank-3 tensor; it is nonzero for a medium with a point group that lacks inversion symmetry.

Third-order nonlinear polarization is described by

$$Q_m(\omega_q, \boldsymbol{r}) = \sum D_3\, \chi^{(3)}_{mnop}(\omega_q\,;\, \omega_1, \omega_2, \omega_3) A_{1n}(\omega_1, \boldsymbol{r}) A_{2o}(\omega_2, \boldsymbol{r})$$

$$\cdot A_{3p}(\omega_3, \boldsymbol{r}) \exp[-i(\boldsymbol{k}_1 + \boldsymbol{k}_2 + \boldsymbol{k}_3)\cdot \boldsymbol{r}] \quad (m, n, o, p = x, y \text{ or } z) \tag{3.5}$$

Here a wave with frequency $\omega_q = \omega_1 + \omega_2 + \omega_3$ is generated from waves with frequencies ω_1, ω_2, and ω_3. The degeneracy D_3 takes the values 1 for three waves of equal frequency (e.g., third harmonic), $3!/2! = 3$ for two waves of equal frequency, and $3! = 6$ for three waves all with distinct frequencies. $\chi^{(3)}$ [m²/V²] is generally a rank-4 tensor and nonzero for crystalline media. Nonzero $\chi^{(3)}$ components for typical crystal structures are shown in Appendix B.1. The real part of $\chi^{(3)}$ relates to optical Kerr effects and the imaginary part relates to two-photon absorption.

The polarization involved in generating phase-conjugate light has a variety of sources, as discussed in Chapter 7. The optical nonlinearities that figure in phenomena related to phase-conjugate light are listed in Table 3.1. Early discoveries include stimulated Brillouin scattering in methane gas (1972) and stimulated Raman scattering, but strictly speaking these do not involve phase conjugate light because they are associated with frequency shifts, say $0.01 - 1$ cm⁻¹ for stimulated Brillouin scattering and $100 - 1000$ cm⁻¹ for stimulated Raman scattering.

Table 3.1 Phase-conjugate light generation and optical nonlinearity

Generating method (physical phenomenon)	Nonlinear optical susceptibility related to phase-conjugate light generation
Stimulated Brillouin scattering	$\chi^{(3)}(\omega \pm \Omega\,;\ \omega, -\omega, \omega \pm \Omega)$
Stimulated Raman scattering	$\chi^{(3)}(\omega \pm \Omega\,;\ \omega, -\omega, \omega \pm \Omega)$
Three-wave mixing	$\chi^{(2)}(\omega\,;\ 2\omega, -\omega)$
Degenerate four-wave mixing (optical Kerr effect)	$\mathrm{Re}\,\chi^{(3)}(\omega\,;\ \omega, \omega, -\omega)$
Degenerate four-wave mixing (photorefractive effect)	Equivalent $\chi^{(3)}$ process (included in Pockels effect $\chi^{(2)}(\omega;\, \omega, 0)$)
Two-photon absorption	$\mathrm{Im}\,\chi^{(3)}(\omega\,;\ \omega, -\omega, \omega)$
Optical echo	$\mathrm{Im}\,\chi^{(3)}(\omega\,;\ \omega, \omega, -\omega)$

Three-wave mixing in a second-order nonlinear optical medium generates phase-conjugate light with no frequency shift.[3] A phase-matching condition must be satisfied. The second harmonic must be supplied as a pump wave. Nonlinear optical media are limited to those that lack central (inversion) symmetry, and since there are other disadvantages, this concept has only historical value.

Degenerate four-wave mixing has been proposed in order to eliminate the above-mentioned disadvantages.[4] Here, two pump waves having the same frequency ω counterpropagate in a medium having a third-order nonlinear coefficient of optical susceptibility. Then, when a third (probe) wave of frequency ω is injected, phase-conjugate light at frequency ω is emitted in the opposite direction to the probe wave. This is currently the principal way of generating phase-conjugate light since it has the following advantages.

(i)   A single light source can be used because the frequencies involved are the same (this is the origin of the term "degenerate"). There is no frequency shift for the same reason.

(ii)   Not only crystals, but also gases, liquids, semiconductors, and organic materials can be used as media since third-order nonlinear polarization exists for all media. There are no restrictions on the media.

(iii)   Since the phase-conjugate wave is automatically the reverse of the probe wave, the conditions on phase matching are much less stringent as long as the pump wave is in a direction opposite to the injected light.

The third-order coefficient of nonlinear susceptibility $\chi^{(3)}$ is mainly used in the optical Kerr effect, and a wide range of media may be used. The optical Kerr effect produces a change in the refractive index proportional to the intensity of the light I [W/m^2]. The refractive index can be written as

$$n = n_0 + n_2 I \tag{3.6a}$$

$$n = n_0 + n_2 \langle E^2 \rangle = n_0 + n_2 A^2/2 \tag{3.6b}$$

The brackets < > represent time averaging. n_2 is called the nonlinear refractive index. The two definitions shown above are often used. There is a conversion relationship between the real part of $\chi^{(3)}$ and n_2

$$n_2 [\mathrm{m^2/W}] = [3/(c\varepsilon_0 n_0^2)] \chi_R^{(3)} [\mathrm{m^2/V^2}] \tag{3.7}$$

where c [m/s] is the speed of light in vacuo. Conversion factors for various systems of units are shown in Appendix A.

Photorefractivity has an effect similar to that of third-order nonlinearity $\chi^{(3)}$. Illumination of a photorefractive material produces an electric field by changing the space-charge distribution. The Pockels effect results in a spatial variation in refractive index (diffraction grating) corresponding to the incident intensity distribution. Degenerate four-wave mixing based on this process can generate phase-conjugate light at relatively low intensity.

When light at a frequency near the atomic resonance frequency is injected, and when light is injected in the neighborhood of a direct semiconductor's band gap, saturation effects arise due to a conspicuous change in the number of carriers in each energy band. Examples of this include the dependence of saturation absorption and refractive index on intensity, as well as gain saturation. The polarization can no longer be represented by a power series in the electric field. For example, atoms and molecules are replaced by a two-level model, and the polarization is treated as an interaction between this model and the electric field.

(2) Wave Equations with Polarization

For a non-magnetic medium, Maxwell's equations and the constitutive relations can be written as follows.

$$\nabla \times \boldsymbol{E} = -\frac{\partial \boldsymbol{B}}{\partial t}, \qquad \nabla \times \boldsymbol{H} = \frac{\partial \boldsymbol{D}}{\partial t} + \boldsymbol{J} \tag{3.8 a,b}$$

$$\operatorname{div} \boldsymbol{D} = \rho, \qquad \operatorname{div} \boldsymbol{B} = 0 \tag{3.8 c,d}$$

where $\boldsymbol{E}$ [V/m] is the electric field, $\boldsymbol{H}$ [A/m] is the magnetic field, $\boldsymbol{D}$ [C/m^2] is the electric flux density, $\boldsymbol{B}$ [Wb/m^2 = Vs/m^2] is the magnetic flux density, and ρ [C/m^3] is the electric charge density. When polarization $\boldsymbol{P}$ [C/m^2] is present,

$$\boldsymbol{D} = \varepsilon_0 \boldsymbol{E} + \boldsymbol{P}, \qquad \boldsymbol{B} = \mu_0 \boldsymbol{H}, \qquad \boldsymbol{J} = \sigma \boldsymbol{E} \tag{3.9 a,b,c}$$

where μ_0 [H/m] is the magnetic permeability of free space and σ [$\Omega^{-1}\mathrm{m}^{-1}$ = A/(Vm)] is the conductivity. The relationship between electric charge ρ and current density $\boldsymbol{J}$ [A/m^2] is expressed by the differential equation

$$\frac{\partial \rho}{\partial t} + \operatorname{div} \boldsymbol{J} = 0 \tag{3.10}$$

This equation is very important to the photorefractive index. In particular, if the polarization is expressed as a sum of linear and nonlinear parts, then

$$\boldsymbol{P} = \varepsilon_0 \chi^{(1)} \boldsymbol{E} + \boldsymbol{P}_{NL} \tag{3.11}$$

If the electric flux density does not vary with time, then the propagation of the light wave can be described by the equation

$$\nabla^2 \boldsymbol{E} - \frac{1}{c^2} \cdot \frac{\partial^2 (\varepsilon \boldsymbol{E})}{\partial t^2} = \mu_0 \sigma \frac{\partial \boldsymbol{E}}{\partial t} + \mu_0 \frac{\partial^2 \boldsymbol{P}_{NL}}{\partial t^2} \tag{3.12}$$

where $\boldsymbol{P}_{NL}$ [C/m^2] is the nonlinear polarization and $\varepsilon \equiv 1 + \chi^{(1)}$ is the electric permittivity of the medium. The first term on the right-hand side of Equation (3.12) contributes to absorption. The electric field can be derived by solving Equations (3.12) and (3.3) simultaneously.

3.2 Four-Wave Mixing in Optical Kerr Media

Four-wave mixing using optical Kerr media has been advancing both theoretically and experimentally. In addition, it has also become the basis for understanding phase-conjugate light generation by other mechanisms. Accordingly, the first objective will be to explain the principles of generating phase-conjugate light based on four-wave mixing. This section deals with the non-resonant case only. Then, a theoretical treatment that depends on the coupled-mode equations will be given as the foundation for the theory of phase-conjugate light generation. In this context, we will explain how pump, probe, and phase-conjugate waves, all of the same frequency, figure in degenerate four-wave mixing. The effects of absorption and pump attenuation will then be discussed, as will features of the process when the pump and probe are slightly detuned (nearly degenerate four-wave mixing). Finally, we will derive the theory of degenerate four-wave mixing on the basis of the diffraction integrals.

(1) Principles of Phase-Conjugate Light Generation

The configuration for degenerate four-wave mixing is shown in Figure 3.1. Two pump waves E_f and E_b of the same frequency ω are injected into a non-linear medium in opposite directions. In addition, a probe wave E_p of the same frequency ω is injected into the same medium from the E_f side. For an isotropic medium, the nonlinear polarization (complex amplitude) of the three superposed light waves can be represented in terms of the respective complex-amplitude vectors A_j as follows[5]

$$\boldsymbol{P}_c^{(3)} = A\left(\boldsymbol{A}_f \cdot \boldsymbol{A}_p{}^*\right)\boldsymbol{A}_b + A\left(\boldsymbol{A}_b \cdot \boldsymbol{A}_p{}^*\right)\boldsymbol{A}_f + 2C\left(\boldsymbol{A}_f \cdot \boldsymbol{A}_b\right)\boldsymbol{A}_p{}^* \qquad (3.13)$$

where, for simplicity, nonlinear polarization that contributes to the newly-generated light is not considered (actually, this is phase-conjugate light, as we will see below).

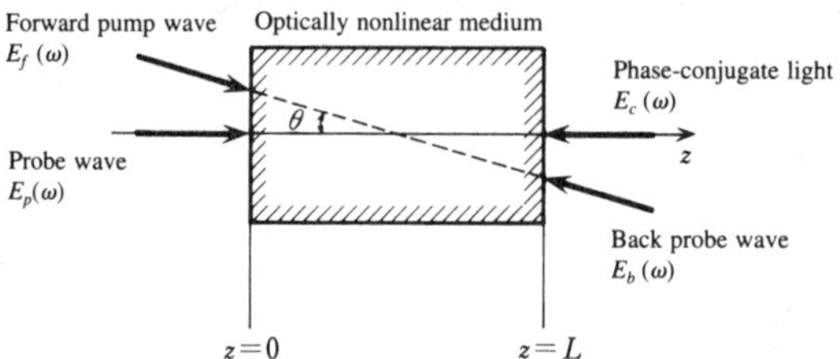

Figure 3.1 Beam arrangement in degenerate four-wave mixing.

ω: wave frequency, θ: angle between probe wave and forward pump wave in the medium.

As Figure 3.2(a) shows, the first term in (3.13) describes a generation process giving rise to conjugate wave E_c ($\propto E_p{}^*$). A stationary three-dimensional diffraction grating is formed by the forward pump E_f and the probe E_p, and the back pump wave E_b experiences Bragg diffraction on this grating. Since E_b is transmitted through the medium, the process is said to involve a transmission grating. The transformation to E_c can be described in terms of a grating vector $\boldsymbol{k}_G = \boldsymbol{k}_f - \boldsymbol{k}_p$. In

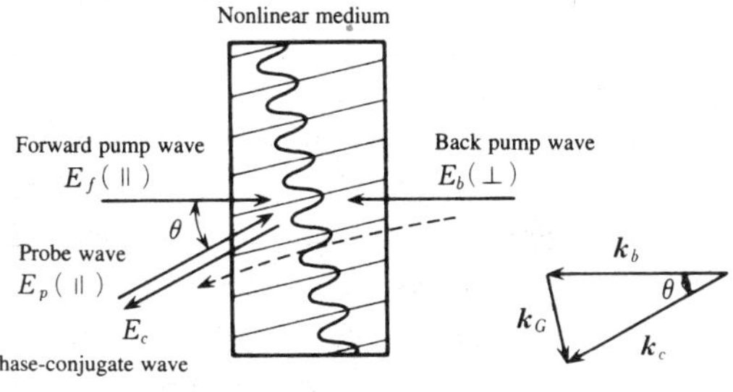

(a) Transmissive diffraction grating

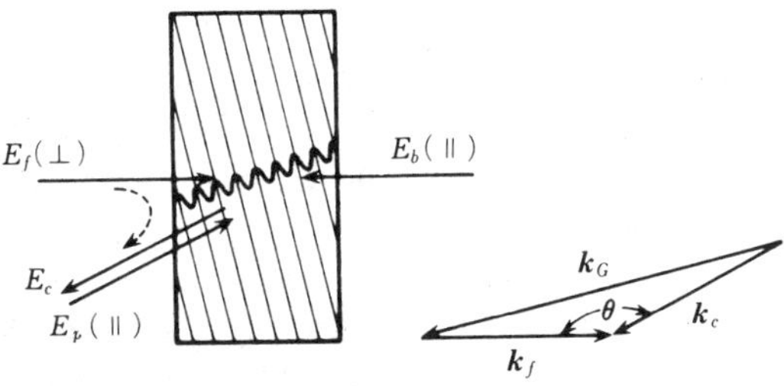

(b) Reflective diffraction grating

Figure 3.2 Two types of spatial diffraction lattices in four-wave mixing.

(a) Phase-conjugate light diffracted in the $k_c = k_b + k_G$ direction, (b) phase-conjugate light diffracted in the $k_c = k_f + k_G$ direction. k_G: diffraction lattice vector; symbols in parentheses indicate the relative direction of polarization for each wave.

a Kerr medium, the interference of E_f and E_p sets up a "population grating," which comprises a spatial distribution of refractive index proportional to intensity (that is, the spatial modulation depends on intensity). The second term in (3.13) describes diffraction of the pump E_f on the grating formed by interference between the pump E_b and the probe, as shown in Figure 3.2(b). Since E_f appears to be reflected from the medium to form E_c, the process is said to involve a reflection grating with grating vector $k_G = k_b - k_p$. If the probe wave and both pump waves are orthogonal, phase-conjugate light is not generated. Now the spacing of the diffraction grating described above is

$$\Lambda = 2\pi/|k_G| = \lambda/[2n\sin(\theta/2)]$$

(3.14)

where θ is the angle the probe wave makes with the pump wave in the medium, and n is the refractive index in the nonlinear medium. If θ is very small, the spacing is wider for a transmission grating than for a reflection grating. Since the two grating vectors $\boldsymbol{k}_f - \boldsymbol{k}_p$ and $\boldsymbol{k}_b - \boldsymbol{k}_p$ are almost orthogonal, one is dominant while the other is relatively weak. In point of fact, only one of them can be expected to dominate.

In contrast, the third term of Equation (3.13) is the probe wave diffracted by the temporal grating formed by the two pump waves modulated at frequency 2ω, thus generating phase-conjugate light. Consequently, phase-conjugate light may also be generated when the waves are orthogonal. The coherences between various levels are related, and can normally be disregarded, except for strong resonances in two-photon transitions (refer to Section 4.3(1)).

The derivation of phase-conjugate light generation due to diffraction gratings takes the same form in degenerative four-wave mixing without the optical Kerr effect. However, when the mechanism forming the diffraction grating changes, a different treatment is required since it becomes necessary for the spatial modulation depth and the phase difference between the interference distribution and the refractive-index diffraction grating to change.

(2) Formulation of the Coupled-Mode Equations

The setup is the same as in Figure 3.1. Let pump waves, a probe wave, and a signal wave (in reality, a phase-conjugate wave, as will be discussed) be plane waves with the same frequency ω. Suppose the pump waves are propagating in opposite directions. Assume further that the optical Kerr medium (having a third-order nonlinear coefficient of optical susceptibility $\chi^{(3)}$ [m²/V²]) occupies the interval $0 < z < L$ along the z axis, which coincides with the direction of the probe wave. The electric field of these four waves is expressed by the following equation.

$$E_j(\boldsymbol{r}, t) = (1/2) A_j(\boldsymbol{r}) \exp[i(\omega_j t - \boldsymbol{k}_j \cdot \boldsymbol{r})] + \text{c.c.} \quad (j = f, b, p, c)$$

$$(3.15)$$

where A_j [V/m] is the slowly varying amplitude, $\boldsymbol{k}_j$ [m⁻¹] is the wave number vector, $\boldsymbol{r}$ [m] is the position vector, and c.c. denotes the complex conjugate. For simplicity, subscripts are used to distinguish among the four light waves here since all the light waves and oblique polarization directions are assumed to be equivalent. Since we are looking for a steady-state solution in this section, A_j does not contain a time dependence.

If optical absorption is disregarded and the optical medium is assumed non-linear and nonmagnetic, the wave equation containing the optical nonlinear process can be expressed by the following equation.

$$\nabla^2 \boldsymbol{E} - \frac{1}{c^2} \cdot \frac{\partial^2 (\varepsilon \boldsymbol{E})}{\partial t^2} = \mu_0 \frac{\partial^2 \boldsymbol{P}_{NL}}{\partial t^2}$$

$$(3.16)$$

where E [V/m] is the total electric field, P_{NL} [C/m^2] is the nonlinear optical polarization, ε is the permittivity tensor, μ_0 [H/m] is the permeability of free space, and c [m/s] is the speed of light in vacuo. A third-order nonlinear polarization is formed by the above-mentioned pump, probe, and phase-conjugate waves.

$$P_{NL} = (1/2)\,\varepsilon_0 \sum_{j,k,l} D_3\,\chi^{(3)}(\omega_i\,;\,\omega_j,\,\omega_k,\,\omega_l)$$

$$\cdot A_j(\omega_j)\,A_k(\omega_k)\,A_l(\omega_l)\exp\{i[(\textstyle\sum\omega_j)\,t-(\sum k_j)\cdot r]\}+\text{c.c.} \tag{3.17}$$

In the above equation, D_3 is the degeneracy, and the subscripts i, j, k, and l are any of f, b, p, or c. The subscripts on the third-order nonlinear susceptibility $\chi^{(3)}$ have been omitted since all electric field polarizations are assumed to be equivalent. Although, in general, the components of frequency $\omega_i = \omega_j \pm \omega_k \pm \omega_l$ are produced from the above-mentioned polarization, since a frequency of ω is also assumed for the generated phase-conjugate light, the relation between the probe and phase-conjugate waves is expressed as follows for an isotropic medium:

$$P_{NL} = (6/2)\,\varepsilon_0\chi^{(3)}A_f A_b A_c{}^*\exp\{i[\omega t-(k_f+k_b-k_c)\cdot r)]\}+\text{c.c.} \tag{3.18a}$$

$$P_{NL} = (6/2)\,\varepsilon_0\chi^{(3)}A_f A_b A_p{}^*\exp\{i[\omega t-(k_f+k_b-k_p)\cdot r)]\}+\text{c.c.} \tag{3.18b}$$

where * denotes the complex conjugate. The other terms contributing to P_{NL} are proportional to $(6A_f\,A_f{}^* + 6A_b\,A_b{}^* + 3A_p\,A_p{}^* + 6A_c\,A_c{}^*)A_p$ and $(6A_f A_f{}^* + 6A_b\,A_b{}^* + 6A_p\,A_p{}^* + 3A_c\,A_c{}^*)A_c$, and are omitted since they are similar to those in (3.18). If the probe wave and the signal light are taken to propagate in opposite directions,[6] then

$$k_f+k_b=0, \qquad k_p+k_c=0 \tag{3.19}$$

Equation (3.19) is called the phase-matching condition. If we substitute this in Equation (3.16) and separate the frequency and wave number terms, we get the following equations.

$$dA_p{}^*/dz = i(3\omega\chi^{(3)}/2cn)(2|A_f|^2+2|A_b|^2+|A_p|^2+2|A_c|^2)A_p{}^*$$
$$+i(3\omega\chi^{(3)}/cn)(A_fA_b)^*A_c \tag{3.20a}$$

$$dA_c/dz = i(3\omega\chi^{(3)}/2cn)(2|A_f|^2+2|A_b|^2+2|A_p|^2+|A_c|^2)A_c$$
$$+i(3\omega\chi^{(3)}/cn)(A_fA_b)A_p{}^* \tag{3.20b}$$

The second derivative can be disregarded in the above equation because $|d^2A_j/dz^2| \ll |k_j(dA_j/dz)|$, assuming that the change in the electric field amplitude per wavelength is gradual. This is called the slowly-varying amplitude approximation (or adiabatic approximation). The pump wave is obtained by replacing subscripts p and f in Equations (3.20) by c and b.

The second term on the right-hand side of (3.20b) represents the leftward-propagating (back) pump wave diffracted on the grating generated by the forward pump and probe waves, thus giving rise to a phase-conjugate wave. The first term on the right-hand side of Equations (3.20a) and (3.20b) represents the optical Kerr effect, which alters only the phase of each electric field, but does not bring about any energy exchange. In addition, since the pump intensity is large compared to the probe and conjugate intensities, the contributions of the latter to the optical Kerr effect are neglected. Accordingly, the substitutions

$$A_p(z) \equiv A_p'(z)\exp[-i(3\omega\chi^{(3)}/cn)(|A_f|^2+|A_b|^2)z] \tag{3.21a}$$

$$A_c(z) \equiv A_c'(z)\exp[i(3\omega\chi^{(3)}/cn)(|A_f|^2+|A_b|^2)z] \tag{3.21b}$$

yield the coupled-mode equations[6]

$$dA_p^*/dz = ixA_c \tag{3.22a}$$

$$dA_c/dz = ix^* A_p^* \tag{3.22b}$$

$$x^* \equiv (3\omega\chi^{(3)}/cn)A_fA_b = (6\pi\chi^{(3)}/n\lambda)A_fA_b \tag{3.23}$$

where x [m^{-1}] is the coupling coefficient, n is the nonlinear index of refraction, and λ is the wavelength. Since there is no risk of confusion, the minus sign is omitted in Equations 3.22. Equations (3.22a) and (3.22b) are the basic equations for finding the probe and phase-conjugate electric fields in degenerate four-wave mixing.

For simplicity, assume the two pump waves remain undepleted in passing through the medium. If the pump intensity is much greater than the probe intensity, and if the conjugate wave is similar in intensity to the probe, that is, $|A_p|, |A_c| \ll |A_f|, |A_b|$, this assumption is adequate as a linear approximation. The coupling coefficient is treated as a constant. Applying boundary conditions on $A_p(0)$ and $A_c(L)$, we obtain the following solution of (3.22):

$$A_p^*(z) = \frac{i|x|}{x^*}\cdot\frac{\sin(|x|z)}{\cos(|x|L)}A_c(L) + \frac{\cos[|x|(z-L)]}{\cos(|x|L)}A_p^*(0) \tag{3.24a}$$

$$A_c(z) = \frac{\cos(|x|z)}{\cos(|x|L)}A_c(L) + \frac{ix^*}{|x|}\cdot\frac{\sin[|x|(z-L)]}{\cos(|x|L)}A_p^*(0) \tag{3.24b}$$

In particular, if $A_p(0) \neq 0$ and $A_c(L) = 0$, that is, only the probe wave is incident, the complex amplitude at the reflection point is

$$A_p^*(L) = A_p^*(0)\sec(|x|L) \tag{3.25a}$$

and we get

$$A_c(0) = -i[(x^*/|x|)\tan(|x|L)]A_p{}^*(0) \tag{3.25b}$$

for the complex amplitude of the phase-conjugate wave. Equation (3.25b) means that, when the probe is injected into a nonlinear optical medium, the complex amplitude of the generated wave is proportional to the complex conjugate of the probe amplitude, and that the generated wave becomes positive phase-conjugate to the probe.

In Equation (3.25b), if we take $\pi/4 < |x|L < 3\pi/4$, then $|A_c(0)| > |A_p(0)|$, and the amplitude of the phase-conjugate wave is larger than that of the probe. This comes about because the conjugate wave takes up energy from the pump and amplifies the probe wave. Moreover, when $|x|L = \pi/2$, $A_c(0)/A_p(0) = \infty$, hinting at oscillation. However, in reality there is no oscillation because of pump depletion due to absorption by the medium and power transfer to the phase-conjugate wave.

Figure 3.3 shows the position dependences of the phase-conjugate and probe intensities as found from Equation (3.24).[6] The intensity of the phase-conjugate wave increases toward the left side of the diagram. On the other hand, the probe intensity increases toward the right side. That is, both the phase-conjugate and the probe waves are amplified.

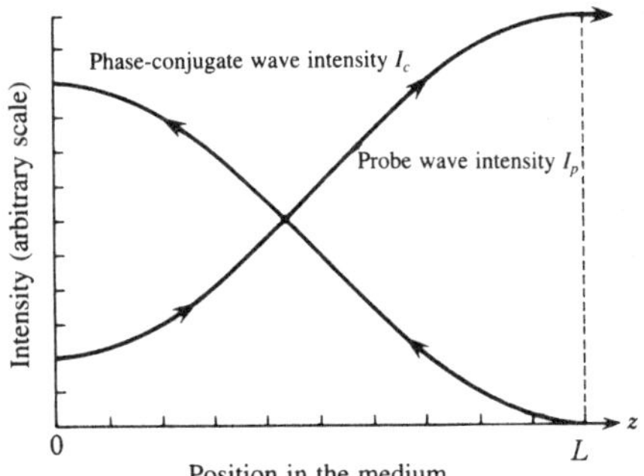

Figure 3.3 Position dependences of conjugate and probe intensities in degenerate four-wave mixing. *(After Yariv)*[6]

Boundary conditions: $A_p(0) \neq 0$, $A_c(L) = 0$, $\pi/4 < |x|L < 3\pi/4$
L = medium length, x = coupling coefficient

The reflectance R (generating effect) is defined as the ratio of the phase-conjugate intensity I_c to the probe intensity I_p

$$R = |A_c(0)|^2/|A_p(0)|^2 \tag{3.26}$$

The transmittance T for the probe wave is defined as

$$T = |A_p(L)|^2 / |A_p(0)|^2 \tag{3.27}$$

From the above we get the following expressions for the reflectance and transmittance:

$$R = \tan^2(|\varkappa| L) \tag{3.28}$$

$$T = \sec^2(|\varkappa| L) \tag{3.29}$$

In particular, when $\varkappa L$ is sufficiently small in comparison with $\pi/4$, the reflectance can be approximated by

$$R \doteqdot (|\varkappa| L)^2 \tag{3.30}$$

Consequently, the reflectance R can be measured as a function of $(I_f I_b)^{1/2}$ (I_f and I_b are the pump intensities). If we extrapolate to small R, we find the third-order coefficient of optical nonlinearity $\chi^{(3)}$ from the coefficient of proportionality. Indeed, many publications cite $\chi^{(3)}$ values calculated in this way.

We now derive the numerical value of the coupling coefficient $\varkappa$. For CS_2 at 694 nm, we have $n_2 = 1.2 \times 10^{-11}$ esu (Table 7.2) and $n_0 = 1.62$. With the conversion factor from Table A (Appendix A), this yields $\chi^{(3)} = 7.2 \times 10^{-21}$ m^2/V^2. If the pump intensities are $I_f = I_b = 1 \times 10^{10}$ W/m^2 (= 1 MW/cm^2), then we get $A_f = A_b = 3.7 \times 10^6$ V/m from Equation (A.5) (Appendix A). When this is substituted in Equation (3.23), we obtain $\varkappa = 1.7$ m^{-1}. When the length of the medium is $L = 1$ cm, we find from Equation (3.30) that the reflectance R becomes 2.9×10^{-4}.

(3) Effects of Absorption

In the previous section we made two assumptions for analytical solutions, namely, (i) the intensities of the two pump waves are sufficiently high compared to the probe and phase-conjugate waves, and (ii) pump depletion is neglected. These assumptions become unsuitable because much of the pump wave energy is transferred to the phase-conjugate wave if the reflectance is large. Accordingly, the following shows the characteristics when the effects of absorption are considered.

The optical arrangement for degenerate four-wave mixing is the same as that in Figure 3.1, and the four light wave frequencies and electric fields are as in Equation (3.15). Here, however, we use wave Equation (3.12) with allowance for absorption by the medium. Both polarization and process are the same as in Section 3.2(2), and the coupled-mode equations for the probe and the phase-conjugate waves are as follows.

$$dA_p{}^* / dz = iKA_c - (\alpha_0/2) A_p{}^* \tag{3.31a}$$

$$dA_c / dz = iK^* A_p{}^* + (\alpha_0/2) A_c \tag{3.31b}$$

$$K^* \equiv (3\omega\chi^{(3)}/nc) A_f(0) A_b(L) \exp(-\alpha_0 L/2) \tag{3.32}$$

where K [m^{-1}] is the complex coupling coefficient with allowance for pump depletion and $\alpha_0 = \sigma\mu_0 c/n$ [m^{-1}] is the small-signal intensity absorption coefficient. When deriving Equations (3.31), we disregard the influence of the probe and conjugate waves and assume that pump depletion is due to absorption alone. In addition, the term for the optical Kerr effect is omitted from Equation (3.20).

Taking $A_p(0) \neq 0$ and $A_c(L) = 0$ as boundary conditions and solving Equations (3.31a) and (3.31b), we get the following for the complex amplitudes of the probe and the phase-conjugate waves at both ends of the medium.

$$A_p{}^*(L) = \frac{BA_p{}^*(0)\exp(\alpha_0 L/2)}{B\cos(BL) + (\alpha_0/2)\sin(BL)} \tag{3.33a}$$

$$A_c(0) = \frac{-iK^* A_p{}^*(0)\tan(BL)}{B + (\alpha_0/2)\tan(BL)} \tag{3.33b}$$

where

$$B = [\,|K|^2 - (\alpha_0/2)^2\,]^{1/2} \tag{3.34}$$

Thus, the reflectance is

$$R = \left| \frac{K^*\tan(BL)}{B + (\alpha_0/2)\tan(BL)} \right|^2 \tag{3.35}$$

If the absorptance is $\alpha_0 = 0$ in Equation (3.35), this naturally reverts to Equation (3.28).

In the low-reflectance limit, in particular, the probe electric field is approximately constant in Equation (3.36b) (Equations (3.36a and b) are obtained from (3.31a) and by the change of variables $A_p{}^*(z) = \exp(-\alpha_0 z/2)A_p{}'^*$ and $A_c(z) = \exp(\alpha_0 z/2)A_c{}'$),

$$dA_p{}'^*/dz = iKA_c{}'\exp(\alpha_0 z) \tag{3.36a}$$

$$dA_c{}'/dz = iK^* A_p{}'^*\exp(-\alpha_0 z) \tag{3.36b}$$

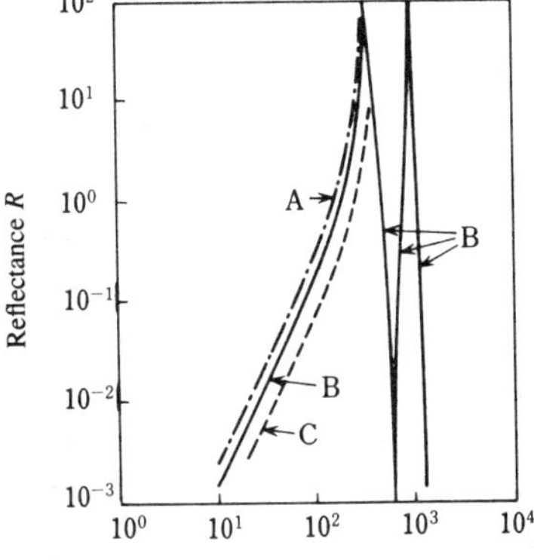

Figure 3.4 Reflectance as a function of coupling coefficient for degenerate four-wave mixing. *(After Caro and Gower © 1982 IEEE)*[7]

A : $\alpha_0 = 0$,   B : $\alpha_0 = 1.2\,\mathrm{cm}^{-1} = 120\,\mathrm{m}^{-1}$,
C : $\alpha_0 = 3\,\mathrm{cm}^{-1}$, $L = 0.5\,\mathrm{cm}$.

so that the reflectance can be written as

$$R \approx \frac{|K|^2}{\alpha_0{}^2}[1-\exp(-\alpha_0 L)]^2$$

$$(3.37)$$

Equation (3.37) has frequently been confirmed by experiment for low-reflectance saturable absorbers.

Figure 3.4 shows the calculated reflectance R with allowance for absorption. Since the figure is obtained from Equation (3.35), values for R at various absorptances α_0 can be shown. When the absorptance α_0 is comparatively small, R is proportional to $|K|^2$ for low R. In addition, R decreases when α_0 increases. Furthermore, when the coupling coefficient increases, the denominator of Equation (3.35) approaches zero, and $R \rightarrow \infty$ (solid curves), showing oscillation. In reality, however, there is no oscillation since power transfer from pump to conjugate and probe waves has been left out of consideration, as has the large optical nonlinearity. At larger $|K|$ values, R is an oscillating function of $|K|$.

A fundamentally different treatment must be given for the case of resonance absorption. The details are covered in Section 4.1.

(4) Effects of Pump Depletion

Next, with Equations (3.20) as a starting point, we examine the situation when pump depletion is considered. Here the invariants in the Manley-Rowe relationship are

$$J_1 = I_b(Z) + I_c(z), \qquad J_2 = I_p(z) + I_c(z), \qquad J_3 = I_f(z) - I_c(z) \qquad (3.38)$$

where the $I_j(z) \equiv |A_j(z)|^2$ are the optical intensities of the waves. Hence, the total energy $I_0 = \Sigma I_j(z)$ is also invariant. When only the optical intensity is looked at, an established, standardized method can be used for the parametric process.[8] Furthermore, the complex amplitude for each light wave can be divided into the real amplitude and phase components

$$A_j(z) = u_j(z)\exp[i\varphi_j(z)]\exp(\mp i\eta I_0 z)$$

$$(3.39)$$

Here $\eta \equiv 3\omega\,\chi^{(3)}/cn$, u_j and φ_j are real numbers; the upper sign is taken for the probe and forward pump waves, and the lower sign is taken for the conjugate and back pump waves. This leads to the following equations from the real and imaginary parts of Equations (3.20 a and b)

$$du_p/dz = -\eta u_f u_b u_c \sin\Phi \qquad (3.40a)$$

$$du_c/dz = \eta u_f u_b u_p \sin\Phi \qquad (3.40b)$$

$$d\Phi/dz = (\eta/2)(u_p{}^2 + u_b{}^2 - u_c{}^2 - u_f{}^2) + \cot\Phi\, d/dz(\ln u_f u_b u_p u_c) \qquad (3.40c)$$

$$\Phi \equiv \varphi_p + \varphi_c - \varphi_f - \varphi_b \qquad (3.40d)$$

Here the invariants from Equations (3.38) have been used. If we assume that the phase-conjugate wave intensity is a function of $y \equiv u_c^2$ alone, we get the following equation from (3.40c) by using the method of variation of the parameter.

$$[y(J_1-y)(J_2-y)(J_3+y)]^{1/2}\cos\Phi + [(J_1+J_2-J_3)y/4 - (y^2/2)] = \Gamma_0$$

(3.41)

Here Γ_0 is a constant. Equation (3.40b) is used to solve the above equation. Equation (3.41) can also be derived by using Lagrangian methods[9] (Appendix C).

Let the boundary conditions be $I_f(z = 0) = I_{f0}$, $I_b(z = L) = I_{bL}$, $I_p(z = 0) = I_{p0}$ and $I_c(z = L) = 0$, and let the intensity of the conjugate wave be subject to $I_c(z = 0) = y_0$, where L is the thickness of the medium. Now we write Equation (3.40b) for the conjugate wave in terms of y and the invariants, obtaining[10]

$$2\eta L = \int_0^{y_0} [yQ(y)]^{-1/2}dy$$

(3.42a)

$$Q(y) \equiv (I_{bL}-y)(I_{p0}+y_0-y)(I_{f0}-y_0+y)$$
$$-y[(I_{bL}-I_{f0}+I_{p0}+2y_0)/4-(y/2)]^2$$

(3.42b)

The solution of Equation (3.42a) can be expressed in Jacobian functions. For example, when $Q(y) = 0$ has three real roots, C_1, C_2, and C_3 ($C_1 > C_2 > 0 > C_3$), the phase-conjugate wave intensity can be expressed as

$$y_0 = \frac{C_2 C_3 \operatorname{sn}^2(l\,|\,m)}{C_3 - C_2 \operatorname{cn}^2(l\,|\,m)}$$

(3.43)

where $l = \eta L[3C_1(C_2 - C_3)/4]^{1/2}$ and $m = C_2(C_1 - C_3)/[C_1(C_2 - C_3)]$.

A numerical plot of the reflectance R is shown in Figure 3.5. The multi-valuedness of the coupling $|\varkappa|L$ corresponding to a given R results from the periodicity of the elliptical functions.[10] This behavior can be explained as follows. When the coupling coefficient increases, more and more power is transferred from the back pump wave I_b to the phase-conjugate wave I_c. Since the sum of I_b and I_c is invariant, I_b, which is the source of I_c, decreases. If the coupling coefficient exceeds a certain level, R falls. This shows that the multi-valuedness is tied to the phase characteristics. $R = 5$ corresponds to the case where all incident I_{bL} is changed to I_c. In addition, when the forward pump intensity I_{f0} and the back pump intensity I_{bL} do not match, I_c is always zero unless there is an incident probe. When I_{f0} and I_{bL} match exactly, limited phase-conjugate power may be generated from optical noise, even when there is no probe input.[9] The dashed curve in the figure represents the solution of Equation (3.28) when pump depletion is ignored (small signal). When the coupling $|\varkappa|L \ll \pi/2$, the solution is exact.

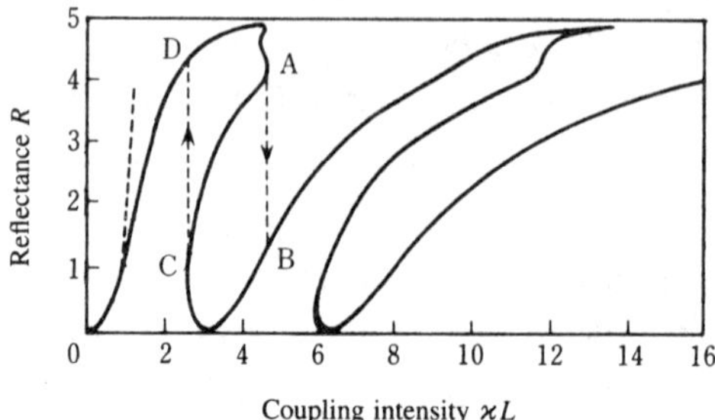

Figure 3.5 Reflectance of an optical Kerr medium with pump wave depletion. *(After Paré, Piché and Bélanger)*[10]

The dashed curves show theoretical small-signal values from Equation (3.28). A, B, C and D are quasi-static variations in probe wave intensity (refer to Section 8.4) $\varkappa \equiv \eta u_f u_b$, L is the thickness of the medium, $I_{f0} = I_{bL}$, $I_{p0} = 0.2\, I_{bL}$.

(5) Nearly Degenerate Four-Wave Mixing

When the probe frequency is different from that of the pump wave, we have nondegenerate four-wave mixing. The three-dimensional diffraction grating formed by the probe and one of the pumps moves through the medium at a constant speed. When the other pump wave is reflected from this grating, there is a frequency shift due to the Doppler effect, and phase-conjugate light at a frequency different from that of the probe is generated. This behavior is not only important in applications, but also very interesting in relation to transient response. In particular, if the amount of the frequency shift is very small, we have nearly degenerate four-wave mixing.

Suppose the frequency ω_p of the probe wave E_p is offset by an amount Ω ($|\Omega|/\omega_0 \ll 1$) from the frequency ω_p of the pump waves E_f and E_b (Figure 3.6). Let the pump waves be injected in opposite directions into a medium with optical nonlinearity $\chi^{(3)}$. The output conjugate wave E_c results from the phase interaction of the waves in the nonlinear medium. The electric fields of the four waves take the same form as that of Equation (3.15). If we assume a constant value for $\chi^{(3)}$ within a narrow frequency range, the nonlinear polarization is the same as that of Equation (3.18), that is,

$$P_{NL} = (6/2)\,\varepsilon_0 \chi^{(3)} A_f A_b A_p{}^* \exp(i\{[\omega_0 + \omega_0 - (\omega_0 + \Omega)]t$$
$$-[(\boldsymbol{k}_f + \boldsymbol{k}_b - \boldsymbol{k}_p)\cdot\boldsymbol{r}]\}) + \text{c.c.} \tag{3.44}$$

We substitute this in wave Equation (3.16) and follow the same procedure as in Section 3.2(2). Depletion of the electric fields of the two pumps can be neglected, and an adiabatic approximation can be employed. Moreover, since

$k_f + k_b = 0$, the phase mismatch becomes minimal in the direction opposite that

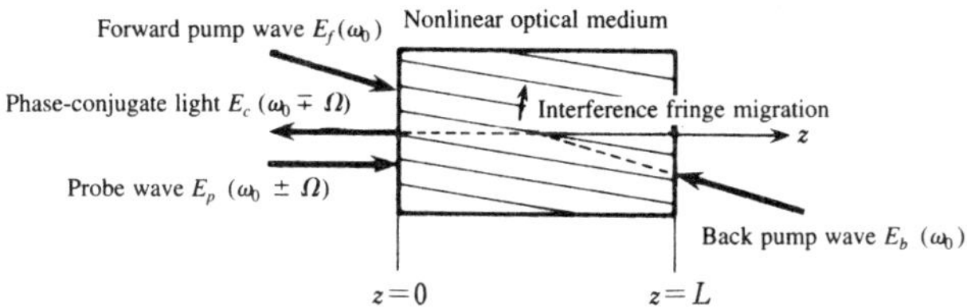

Figure 3.6 Production of phase-conjugate light by nearly-degenerate four-wave mixing.

Because of the frequency difference between E_f and E_p, the interference fringes (spatial diffraction grating) migrate at a constant speed. E_b is again diffracted on the grating to form E_c, but the frequency is shifted by the Doppler effect. Here ω_0 is the pump frequency and Ω is the frequency offset. The frequency offsets in the probe and conjugate waves take opposite signs. The direction of migration of the interference fringes as indicated by the arrow corresponds to the upper sign.

of the probe, so that the output wave E_c is generated in this direction. If the probe frequency is $\omega_0 + \Omega$, E_b is diffracted on a grating (fringes) moving away from it, and the output frequency is downshifted to $\omega_0 - \Omega$. Consequently, we get the following coupled-mode equations.[11]

$$dA_p{}^*/dz = ix_p A_c \exp(-i\,|\,\delta k\,|\,z) \tag{3.45a}$$

$$dA_c/dz = ix_c{}^* A_p{}^* \exp(i\,|\,\delta k\,|\,z) \tag{3.45b}$$

where x_j is the complex coupling coefficient and δk is the phase mismatch defined as

$$\chi_j{}^* = (3\omega_j \chi^{(3)}/nc)\,A_f A_b \quad (j = p, c) \tag{3.46a}$$

$$\delta k = k_f + k_b - k_p - k_c, \quad |\,\delta k\,| = 2n\Omega/c \tag{3.46b}$$

where n is the index of refraction of the medium and c is the speed of light in vacuo. Equations (3.45a) and (3.45b) are the basic equations for examining the characteristics of the probe wave and the phase-conjugate light when there is detuning between the probe and the pumps.

If we solve Equations (3.45a) and (3.45b) when the pump electric field is constant and when the boundary conditions on the probe are $A_p\,(z = 0) \neq 0$ and $A_c\,(z = L) = 0$, we get the following equations for the probe and the phase-conjugate electric fields:

$$A_p{}^*(L) = \frac{\beta A_p{}^*(0)\exp(-i\,|\,\delta k\,|\,L/2)}{\beta\cos(\beta L) - (i\,|\,\delta k\,|/2)\sin(\beta L)} \tag{3.47a}$$

$$A_c(0) = \frac{-ix_c{}^* A_p{}^*(0)\tan(\beta L)}{\beta - (i\,|\,\delta k\,|/2)\tan(\beta L)} \tag{3.47b}$$

where

$$\beta \equiv [\varkappa_p \varkappa_c^* + (|\delta \boldsymbol{k}|/2)^2]^{1/2} \tag{3.48}$$

We assume here that the detuning is very small and that, if $\varkappa_p \approx \varkappa_c \equiv \varkappa$, the reflectance is given by

$$R = \frac{|\varkappa L|^2 \tan^2(\beta L)}{|\varkappa L|^2 + (|\delta \boldsymbol{k}|L/2)^2 \sec^2(\beta L)} \tag{3.49}$$

Equation (3.49) gives the reflectance when there is detuning between the pump and the probe. If $\Omega = 0$ in this equation, it reverts back to Equation (3.28).

Figure 3.7 shows how the reflectance depends on the detuning.[11] The horizontal coordinate is $\Psi = |\delta k|L/2\pi = [(\delta\lambda/2)(2nL/\lambda^2)]$, the normalized detuning. The reflectance dramatically falls off if there is detuning, and we get steep detuning characteristics when $|\varkappa|L$ is in the neighborhood of $\pi/2$. If the interaction length L is greater than the coherence length determined by the phase mismatch $L_c = \pi/|\delta k|$, then the reflectance for the output wave is dramatically reduced. This is the transient response characteristic relationship

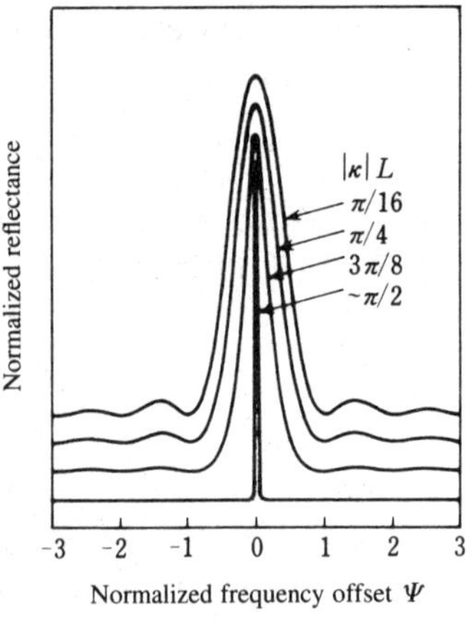

Figure 3.7 Reflectance of phase-conjugate light versus frequency offset. *(After Pepper and Abrams)*[11]

$$\Psi = \frac{\delta k \cdot L}{2\pi} = \frac{\delta\lambda}{2} \cdot \frac{2nL}{\lambda^2}$$

The vertical axis is normalized at $\Psi = 0$. When $L = 1$ cm, $\lambda = 0.5$ μm, and $n = 1.62$ (for CS_2), $\Psi = 1$ on the horizontal axis corresponds to $\delta\lambda/2 = 0.772$ nm.

to be addressed in Section 4.4, which also relates to the optical bandpass filter application to be discussed in Section 8.5.

(6) Degenerate Four-Wave Mixing Theory Based on Diffraction Integrals

We now take up elementary four-wave mixing theory.[4] Although it is not yet as widely accepted as the coupled-mode theory, we will touch upon it since there are so many approaches to it. Figure 3.8 illustrates the coordinate system. Suppose that intense pump waves E_f and E_b (monochromatic plane waves with frequencies $\omega_f = \omega_b = \omega_0$) are injected into an optically nonlinear medium from

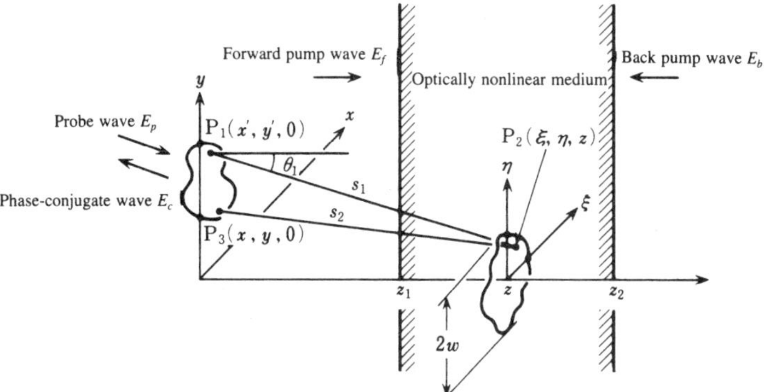

Figure 3.8 Coordinate system for derivation of degenerate four-wave mixing theory from diffraction analysis.

The nonlinear optical medium lies in the region $z_1 < z < z_2$; the pump waves are injected into the medium in opposite directions. P_1 is the position of the probe wave, P_2 a test point in the medium, and P_3 the position of the output wave due to polarization. s_1 is the distance between P_1 and P_2, s_2 the distance between P_2 and P_3, and $2w$ the beam diameter in the (ξ, η) plane of the probe wave.

opposite directions. For convenience, take the propagation of the forward pump wave to be in the z direction, and assume that the nonlinear medium is located in the region $z_1 < z < z_2$. Further assume that the probe wave E_p (frequency ω_p) is injected from the left. The electric field component is described by

$$E_{jm} = (1/2)\,A_{jm}(r)\exp[i(\omega_j t - k_j z)] + \text{c.c.}$$

$$(j = f, b, p;\ \ m = x, y \text{ or } z) \tag{3.50}$$

where $A_{jm}(r)$ is the complex amplitude and k_j is the wave number. Assume that the probe's amplitude A_{pm} is already known at every point in space.

The electric field at point $P_2(\xi, \eta, z)$ due to the probe from point $P_1(x', y', 0)$ can be expressed as follows with the Fresnel-Kirchhoff diffraction integral.

$$A_{pm}(\xi, \eta, z) = ik_p \iint dx'\,dy'\,A_{pm}(x', y', 0)\,K\exp(ik_p s_1)/(2\pi s_1) \tag{3.51}$$

where $k_p = n_p \omega_p / c$ is the probe wave number, n_p is the refractive index at frequency ω_p, c is the speed of light in vacuo, $K = 1 + \cos\theta_1$ is the slope factor, and θ_1 is the angle between s_1 and the z axis. The distance s_1 between P_1 and P_2 is assumed to be large compared to the wavelength. If it is assumed that a third-order optically nonlinear process brings about polarization in the medium, the nonlinear polarization at a frequency $\omega_c = 2\omega_0 - \omega_p$ can be expressed as

$$P_{NLl} = X_{lo}A_{po}{}^* + \text{c.c.} \tag{3.52a}$$

$$X_{lo} = 6\chi^{(3)}_{lmno}(\omega_c;\ \omega_0, \omega_0, -\omega_p)\,A_{fm}A_{bn} \tag{3.52b}$$

where $\chi^{(3)}$ is the third-order coefficient of nonlinear optical susceptibility. The polarization obtained by substituting Equation (3.51) in Equations (3.52) generates a new wave.

This nonlinear polarization sets up the following electric field at a point $P_3(x, y, 0)$ lying outside the medium

$$A_{cl}(x, y, 0) = k_c^2 \int_{z_1}^{z_2} dz \iint d\xi d\eta \, QX_{lo}A_{po}{}^*(\xi, \eta, z)\exp(ik_c s_2)/s_2$$
(3.53)

where $k_c = n_c \omega_c/c$ is the wave number of the output wave, n_c is the refractive index at frequency ω_c, s_2 is the distance between points P_2 and P_3, and Q is the operator representing projection into a surface perpendicular to the polarization vector of the incident probe wave. Equations (3.52 a and b) are substituted in Equation (3.53) and integrated. However, in calculating the phase term the beam radius in the (ξ, η) plane is assumed to be w and the probe beam is assumed narrow. If it is assumed that $z \gg w^2(k_c - k_p)$ and $z \gg k_p w^4/z^2$, the higher-order terms in ξ and η can be neglected. In addition, K and $(s_1 s_2)$ are approximated by the average value of w within its domain. If such an approximation is made, the integral over ξ and η can be replaced by a δ function. Consequently, the output electric field at P_3 due to nonlinear polarization is

$$A_{cl}(x, y, 0)$$

$$= 2\pi i \frac{k_c^2}{k_p} \int_{z_1}^{z_2} dz \exp[i(k_c - k_p)z]\xi(z)\, QX_{lo}A_{po}{}^*\left(\frac{k_c}{k_p}x, \frac{k_c}{k_p}y, 0\right)$$
(3.54)

where $\xi(z) \equiv \bar{K}z^2/(\bar{s}_1, \bar{s}_2)$ and the bar denotes average value. In the above equation, it is assumed that although the two pump intensities depended on z, they did not depend on ξ and η.

Equation (3.54) has the following interpretation. (i) If $k_p = k_c$, the new electric field $A_{c\,m}(r)$ due to the nonlinear polarization is in the same spatial position as, and proportional to the complex conjugate of, the probe field $A_{pm}(r)$. (ii) If $k_p \neq k_c$, $A_{cm}(r)$ is proportional to $A_{pm}{}^*(r)$ in the spatial neighborhood. In short, if $k_p = k_c$ in the case of counterpropagating pump waves, there is phase-conjugate light without phase matching. In order for k_p to be equal to k_c, the pump frequency ω_0 and the probe frequency ω_p must agree, so that the frequency of the phase-conjugate wave ω_c will agree with ω_p ($= \omega_0$). If $k_p \neq k_c$, the conjugate wave — as opposed to the probe — will shift position or be amplified or attenuated.

3.3 Four-Wave Mixing in a Photorefractive Medium

A diffraction grating is formed in media such as $LiNbO_3$, $BaTiO_3$, and $Bi_{12}SiO_{20}$ by illumination, and phase-conjugate light is created through a diffraction process similar to holography. This process can be explained in form by Figure 3.2. However, the physical phenomenon that leads to the formation of this diffraction grating is the photorefractive effect. A band-transport model and a hopping model have been proposed for carrier movement. In this section, after first outlining both models, we will treat the generation of phase-conjugate light based on the band-transport model.

(1) Band-Transport and Hopping Models

Figure 3.9 illustrates the band-transport and hopping models. In the first, a majority carrier (electron) is raised into the conduction band, drifts and diffuses through the band, and finally drops to a donor site left vacant by recombination. A charge distribution corresponding to the intensity distribution is thought to migrate as a unit.[12] In the hopping model, on the other hand, an electron on absorbing a quantum of energy has a constant probability of hopping to adjacent donor sites. A carrier distribution matching the intensity distribution is thought to be formed in the medium.[13] The electron transition and recombination times are short in comparison with other optical processes. The two models are in agreement when the modulation of the charge distribution is low.[14,15] In what follows, the generation of phase-conjugate light is explained on the basis of a band-transport model. If the electric field distribution is described in terms of the charge distribution, both models can be treated similarly.

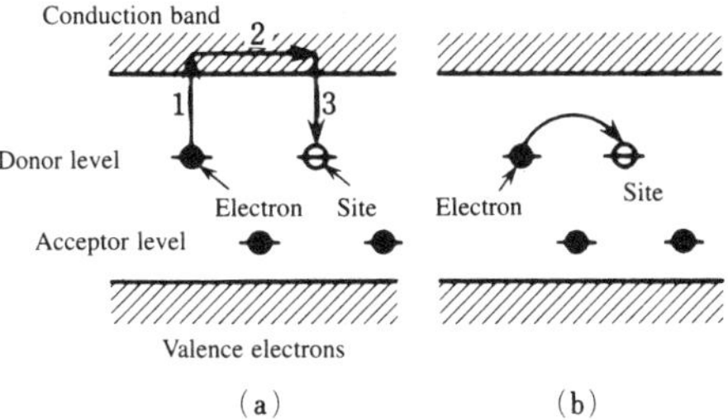

Figure 3.9 Band-transport and hopping models in a photorefractive medium (the carriers are electrons).

(a) Band-transport model. Electrons at the donor level absorb light energy and move to the conduction band (1: optical ionization), electrons migrate through drift and diffusion (2) in the conduction band, and move to a new site by recombination (3).
(b) Hopping model. Electrons at the donor level absorb light energy and hop to adjacent sites with a constant probability.

(2) A Qualitative Explanation of Phase-Conjugate Light Generation Based on the Band-Transport Model

The band-transport mechanism of phase-conjugate light production can be broken down into five steps, as shown in Figure 3.10.[12] The majority carriers are assumed to be electrons.

(i) First, the interference of two waves (one pump wave and the probe in this instance) produces fringes, that is, periodic variations of $I(x)$ in the medium. (For simplicity, the medium is treated as one-dimensional.)

(ii) In conformance with the wave intensity distribution, electrons from the donor level are removed by ionization and migrate to the conduction band. If the electrons are dispersed by scattering, and if there is an externally-impressed electric field, the electrons drift. As a result, the electric charge distribution $\rho(x)$ is created.

(iii) This electric field distribution satisfies the Poisson equation and sets up a steady electric field $E_{st}(x)$.

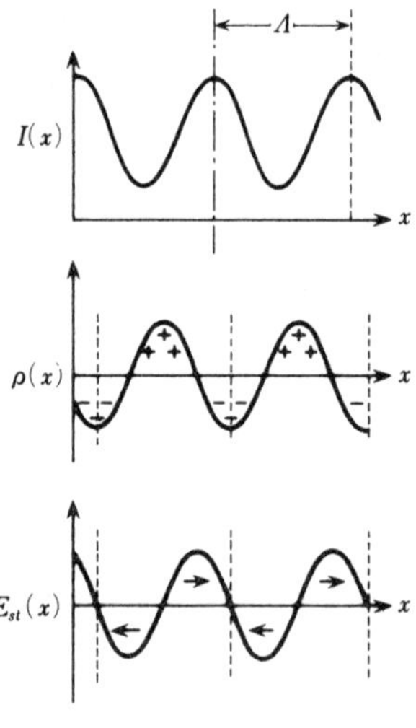

Figure 3.10 Formation of refraction-diffraction grating in a photo-refractive medium (band-transport model).

$I(x)$:   Optical intensity distribution
$\rho(x)$:   Space charge distribution
$E_{st}(x)$: Steady electric field distribution
$\delta n(x)$: Change in refractive index
Λ :   Grating constant
ψ :   Phase difference between $I(x)$ and $\delta n(x)$

(iv) The refractive index shows a change. $\delta n(x)$ varies due to the Pockels effect (first-order electro-optic effect), creating the diffraction grating. In general, there is a phase difference ψ between $I(x)$ and $\delta n(x)$. The value of ψ depends on the grating formation mechanism.

(v) Next, if the reading wave (in this case the other pump wave) is injected into the nonlinear optical medium, there is polarization corresponding to the diffraction grating, and a new phase-conjugate wave is generated.

These phenomena can be described on the basis of the constitutive relations.

(3) Fundamental Equations of the Photorefractive Effect

The constitutive relations follow from the law of conservation of electric charge (continuity), the rate equations for ions, and the Poisson equations. If we assume ρ [C/m^3] as the electric charge density and J [A/m^2] as the current distribution, then we get

$$\frac{\partial \rho(r, t)}{\partial t} + \text{div} \, J(r, t) = 0 \tag{3.55}$$

as the law of conservation of electric charge.

If the majority carriers are electrons or holes, and if we represent the carrier density by n_c [m^{-3}], the ionized donor density by N_D^+ [m^{-3}], the acceptor density by N_A [m^{-3}], and the electric charge of an electron by e [C], then the charge density can be expressed by

$$\rho = \begin{cases} -e(n_c + N_A - N_D^+) & \text{when the majority carriers are electrons} \tag{3.56a} \\ e(n_c - N_A + N_D^+) & \text{when the majority carriers are holes} \tag{3.56b} \end{cases}$$

In the complex expression below, the upper (lower) sign applies when the carriers are electrons (holes). We get the following as the electric current equation.

$$J = e\mu n_c E_{st} \pm k_B T \mu \, \text{grad} \, n_c + pI \tag{3.57}$$

The first term on the right-hand side of the equation above represents the drift current, the second term represents the scattering current, and the third term represents the current produced by the optical electromotive force. Here, μ [m^2/(Vs)] is the migration rate, E_{st} [V/m] is the vector sum of the electric field (which varies much more slowly than the light frequency) and the externally impressed electric field (steady field), k_B [J/K] is Boltzmann's constant, T [K] is the absolute temperature, p [As/J] is an optical emf constant, and I [J/(sm^2)] is the intensity.

Next we consider the rate equation for the ions. If N_D [m^{-3}] is the donor density, then $N_D - N_D^+$ is the number of carriers contributing to optical excitation in the case of electrons or the number of carriers contributing to recombination in the case of holes. Consequently, if s [m^2/J] is the excitation cross section (if this is multiplied by the wave energy $\hbar\omega$, it becomes the areal dimension), and if γ_R [m^3/s] is the coefficient of recombination, then we get

$$\frac{\partial N_D^+}{\partial t} = \begin{cases} (sI+\beta)\,(N_D-N_D^+) - \gamma_R n_c N_D^+ & :\text{electrons} \\ -(sI+\beta)\,N_D^+ + \gamma_R n_c\,(N_D-N_D^+) & :\text{holes} \end{cases} \tag{3.58}$$

as the rate equation for carriers. $\gamma_R n_c N_D^+$ in the above equation expresses the recombination rate. β [s^{-1}] is the thermo-excitation rate.

The Poisson equation is obtained from

$$\text{div}\,\boldsymbol{E}_{st}\,(\boldsymbol{r},\,t) = \frac{\rho(\boldsymbol{r},\,t)}{\varepsilon_s \varepsilon_0} \tag{3.59}$$

where ε_s is the static permittivity and ε_0 [F/m] is the permittivity of free space. If the scalar potential ϕ is defined as

$$\boldsymbol{E}_{st} = -\text{grad}\,\phi \tag{3.60}$$

the Poisson Equation (3.59) can be written as

$$\nabla^2\phi = -\frac{\rho(\boldsymbol{r},\,t)}{\varepsilon_s \varepsilon_0} \tag{3.61}$$

If the Pockels effect (variation in the refractive index proportional to the electric field) is expressed in terms of the relative permittivity, we have

$$\varepsilon_\omega = \varepsilon_n - \varepsilon_n(\vec{R}\boldsymbol{E}_{st})\,\varepsilon_n \tag{3.62}$$

where ε_ω is the permittivity in the optical domain, ε_n is the permittivity in the absence of an electric field, and $\vec{R}$ [m/V] is the rank-3 electro-optic tensor. Electro-optic tensors for typical crystal structures are shown in Table B.2 of the appendix. If E represents the electric field, then the behavior of light waves can be described by the wave equation

$$\nabla^2\boldsymbol{E} - \frac{1}{c^2} \cdot \frac{\partial^2}{\partial t^2}(\varepsilon_\omega \boldsymbol{E}) = 0 \tag{3.63}$$

where c is the speed of light in vacuo. The photorefractive effect for the band-transport model can be solved for by using the above equations. The second term on the right-hand side of Equation (3.62) corresponds to the optically nonlinear polarization term in Equation (3.16).

(4) Steady-State Analysis of Degenerate Four-Wave Mixing in Accordance with the Band-Transport Model

In order to make the contents of this section more easily understandable, we will deal only with electrons as carriers. This type of optical arrangement is the same as that presented for optical Kerr media in Figure 3.1. For simplicity, assume two counterpropagating pump waves (plane waves) E_f and E_b having the same polarization and the same frequency ω. Assume also that the light is injected into a photorefractive medium. Moreover, assume that a phase-conjugate wave E_c is produced if a probe wave E_p of the same frequency ω is injected into the medium. Take the direction of the probe along the z axis, and assume that $0 < z < L$ for the medium (where the optical tensor $\vec{R}$ has rank three). The electric fields of the four waves are expressed as follows.

$$E_j(\boldsymbol{r}, t) = (1/2) A_j(\boldsymbol{r}) \exp[i(\omega t - \boldsymbol{k}_j \cdot \boldsymbol{r})] + \text{c.c.} \quad (j = f, b, p, c) \tag{3.64}$$

where A_j [V/m] is the complex envelope of the slowly varying field, $\boldsymbol{k}_j$ [m^{-1}] is the wave number vector, and $\boldsymbol{r}$ [m] is the position vector.

As previously discussed, a spatial diffraction grating can be of transmission or reflection type. Since we can anticipate that one of them will be weaker, we will assume in the discussion below that a transmission grating is formed. The optical intensity distribution I [W/m^2] formed within the medium by the four waves mentioned above can be expressed by the following equation in the case of a transmission grating.

$$I = (nc\varepsilon_0/2) \left| \sum_j A_j \right|^2$$

$$= I_0 + I_2(z) \exp(-ik_G x) + \text{c.c.} \tag{3.65a}$$

$$I_0 = (nc\varepsilon_0/2) \sum_j |A_j|^2 \tag{3.65b}$$

$$I_2 = (nc\varepsilon_0/2) (A_f A_p{}^* + A_b{}^* A_c) \tag{3.65c}$$

where I_0 is the sum of the intensities, I_2 is the interference term, $k_G = 2k_m \sin(\theta/2) = 2\pi/\Lambda$ is the grating wave number, k_m is the optical wave number in the medium, and Λ is the spacing of the diffraction grating. As mentioned above, although the electric charge distribution and the electric field distribution are formed by the interference fringes, we are assuming that the spatial frequencies of the carrier density n_c and the steady field distribution E_{st} are equivalent to that of the interference fringes. Therefore,

$$n_c = n_{c0} + n_{c2}(z) \exp(-ik_G x) + \text{c.c.} \tag{3.66}$$

$$E_{st} = E_{s0} + E_{s2}(z) \exp(-ik_G x) + \text{c.c.} \tag{3.67}$$

If we substitute these expressions in Equations (3.55) – (3.58) and use the approximations $n_{c0} \ll N_A \ll N_D$ and $\beta \ll sI_0$ (that is, the carrier density due to optical excitation is much higher than that due to thermal excitation), then the steady-state solution for the spatially varying components in the steady field distributions can be written as[12,16]

$$E_{s2} = (I_2/I_0) E_m \exp(i\psi) \tag{3.68}$$

$$E_m = -E_q \left[\frac{E_0^2 + E_T^2}{E_0^2 + (E_q + E_T)^2} \right]^{1/2} \tag{3.69a}$$

$$\tan \psi = \frac{E_0^2 + E_T(E_q + E_T)}{E_0 E_q} \tag{3.69b}$$

where E_m [V/m] is the magnitude of the electric field, $E_q \equiv eN_A/(\varepsilon_s \varepsilon_0 k_G)$ [V/m] is the maximum value of the electric field, $E_T \equiv k_B T k_G/e$ [V/m] is the scattering electric field, E_0 is the externally impressed dc electric field, and ψ is the phase difference between the optical intensity distribution and the electric field.

Let us now consider the magnitude of the electric field. When there is no externally-impressed field E_0 and $E_T < E_q$, we get $E_m \approx -E_T$ and $\psi = \pi/2$. This means that the electric field is governed mainly by the scattering field when the externally impressed field is small. On the other hand, if we take $E_m \approx -E_0$ when $E_T < E_0 < E_q$, the drift electric field has an effect due to the externally-impressed electric field. ψ approaches 0 at this time. The phase difference ψ differs from $\pi/2$ because carriers excited by optical energy are moving in the electric field. As a result of this motion, the point of carrier excitation does not coincide with the point where the carrier is accelerated by the field; the reflectance accordingly drops. If $E_q < E_0$ or $E_q < E_T$, then $E_m \approx -E_q$. The strength of the electric field thus depends on the relationship between the scattering and drift fields.[17] The wave number k_G of the diffraction grating is contained in E_q and E_T and is inversely proportional to the grating spacing Λ. Consequently, the magnitude of the electric field E_m and the phase difference ψ depend on the spacing Λ and the angle formed by the pump and the probe waves.

Now we can find the change in the permittivity (that is, in the refractive index) by substituting Equation (3.68) in Equation (3.62). In order to find the equation for the electric field, we substitute the change in permittivity in wave Equation (3.63) and make use of the slowly-varying-envelope approximation as in the optical Kerr effect. Furthermore, if we pick out only the terms that satisfy $\mathbf{k}_f + \mathbf{k}_b = \mathbf{k}_p + \mathbf{k}_c$ (the phase-matching condition), we get the following coupled-mode equations.

$$dA_p{}^*/dz = (\gamma/I_0)(nc\varepsilon_0/2)(A_f A_p{}^* + A_b{}^* A_c) A_f{}^* \tag{3.70a}$$

$$dA_c/dz = (\gamma/I_0)(nc\varepsilon_0/2)(A_f A_p{}^* + A_b{}^* A_c) A_b \tag{3.70b}$$

$$dA_f/dz = -(\gamma/I_0)(nc\varepsilon_0/2)(A_f A_p{}^* + A_b{}^* A_c) A_p \tag{3.70c}$$

$$dA_b{}^*/dz = -(\gamma/I_0)(nc\varepsilon_0/2)(A_f A_p{}^* + A_b{}^* A_c) A_c{}^* \tag{3.70d}$$

$$\gamma = [i\omega r_{\mathrm{eff}} n^3 E_m \exp(i\psi)]/(2c\cos\theta)$$

$$\equiv ia\exp(i\psi) \tag{3.71}$$

Equation (3.70b) describes the diffraction of the back pump wave on the diffraction grating formed by the forward pump and probe wave, giving rise to phase-conjugate light. Here γ is the complex coupling coefficient, a is positive, and ψ is taken to be a real number. r_{eff} [m/V] is the effective electro-optic coefficient of an anisotropic medium, defined as

$$r_{\mathrm{eff}} = [\varepsilon_n(\vec{R}E_{st})\varepsilon_n]_{ij}/n_0{}^3 n_B E_{st} \tag{3.72}$$

Here n_B is the index of refraction for the ordinary ray n_0 or the extraordinary ray n_e depending on the interaction of the waves.

If the intensities of the probe and phase-conjugate waves are sufficiently small relative to the pump and the coupling is weak, we see from Equations (3.70c) and (3.70d) that the pump amplitude can be assumed not to vary during propagation through the medium. If Equations (3.70a) and (3.70b) are solved subject to the boundary conditions $A_p(0) \neq 0$ and $A_c(L) = 0$, we get the following equations for the complex amplitudes of the probe and conjugate waves[18]

$$A_p{}^*(L) = A_p{}^*(0)\frac{1}{r^{-1}\exp(-\gamma L)+1}(r^{-1}+1) \tag{3.73a}$$

$$A_c(0) = A_p{}^*(0)\frac{A_f/A_b{}^*}{r^{-1}\exp(-\gamma L)+1}[\exp(-\gamma L)-1] \tag{3.73b}$$

where $r \equiv I_b/I_f$ is the optical intensity ratio of the pump waves.

The reflectance R^- is

$$R^- = \left|\frac{\sinh(\gamma L/2)}{\cosh\{(\gamma L/2)+[(\ln r)/2]\}}\right|^2 \tag{3.74}$$

and the probe transmittance T^- is

$$T^- = \left|\frac{\cosh[(\ln r)/2]\exp(\gamma L/2)}{\cosh\{(\gamma L/2)+[(\ln r)/2]\}}\right|^2 \tag{3.75}$$

Since Equations (3.74) and (3.75) contain only r and γL, the reflectance and the transmittance are governed not by the optical intensity itself, but by the intensity ratio. When $r = \exp(aL\sin\psi)$, the maximum reflectance becomes

$$R_{\max} = \left| \frac{\sinh[(iaL/2)\exp(i\psi)]}{\cos[(aL/2)\cos\psi]} \right|^2 \tag{3.76}$$

In particular, when there is no externally impressed electric field, $R_{\max} = \sinh^2$ $(aL/2)$ when $\psi = \pi/2$.

Although the above results are for transmission gratings, we would get the same results for reflection gratings if we replaced the electric field in Equation (3.65c) with $(A_f^*A_c + A_b A_p^*)$.

(5) Interpretation via Phase-Conjugate Wave Reflectance (Feedback)

The efficiency of conversion of probe to conjugate-wave energy by refractive effects depends on the phase difference ψ between the optical intensity distribution and the refractive-index distribution. In order to understand this argument easily, let

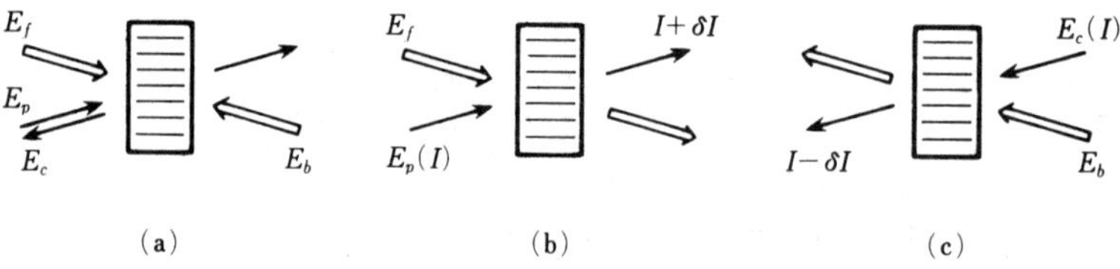

Figure 3.11 Two-photon coupling analysis of four-wave mixing.

(a) Four-wave mixing, (b) Two-photon coupling via forward pump and probe waves, (c) Two-photon coupling via back pump and phase-conjugate waves. The direction of energy transport depends on the phase difference ψ between the optical intensity distribution and the refractive-index distribution.

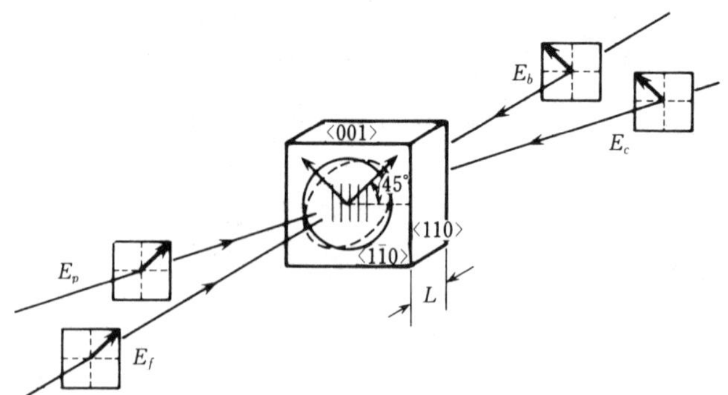

Figure 3.12 Improvement in the reflectance by positive feedback for cubic crystals.

The circle (solid curve) and the ellipse (dashed curve) stand for the index ellipsoid when there is no steady electric field E_{st} and when $E_{st} \parallel k_G$. In the latter case the major axis is at $\pm 45°$ to the plane of incidence.

it be restricted to a transmission grating. We will analyze two-photon mixing for a qualitative consideration of four-wave mixing, as shown in Figure 3.11. In Figure 3.11(b), interference fringes are created by the forward pump wave E_f and probe wave E_p. When $\psi = \pi/2$, a stable diffraction grating formed by the two waves exists since there is no change in the relative phase difference between E_f and E_p. Consequently, when $\psi = \pi/2$, the phase is the same (phase difference = 0) as that of one of the waves propagating through the medium, and the amplitude is increased, but the other wave has opposite phase (phase difference = π), and the amplitude is decreased. As a result, energy transfer from E_f to E_p[17,19] is accomplished efficiently. That is, E_f self-diffracts, and the probe E_p is amplified. On the other hand, as shown in (c), the same refractive grating as in (b) is formed by the back pump E_b and the phase-conjugate wave E_c, but since the direction of propagation is the opposite of (b), the phase difference ψ reverses its sign, and the energy of the phase-conjugate wave E_c depletes as it propagates. Therefore, when the system is looked at from the viewpoint of phase-conjugate light generation, we can see a system of positive feedback.[20,21] As a result, the signs of the subscripts in Equations (3.74) and (3.75) are reversed.

There are structurally anisotropic crystals which, in general, have two special polarization conditions not subject to polarization changes along the propagation path.[22] In particular, consider the cubic crystal structures (such as in BSO, GaAs, and InP) often present in refractive media. Where such a structure is set up, as shown in Figure 3.12, the special polarization conditions found from the index ellipsoid represent ±45° polarization to the plane of incidence. Accordingly, each wave is linearly polarized at ±45°. If they are incident, the gratings formed by E_f and E_p and by E_b and E_c function independently. In this case, the phase difference for the above-mentioned ±45° linear polarization diffracted by the same grating becomes π. Consequently, if an appropriate case in Figure 3.11(b) is $\psi = \pi/2$, a suitable case for Figure 3.11(c) would be $\psi = -\pi/2 + \pi = \pi/2$, and both the probe and phase-conjugate waves are amplified as they propagate. Therefore, there is positive feedback for the whole system, and phase-conjugate light is generated efficiently. We get the following for the reflectance when pump depletion is neglected.[21,23]

$$R^+ = \left| \frac{\sinh \dfrac{(1-r)\,\gamma L}{(1+r)2}}{\sinh\left[\dfrac{(1-r)\,\gamma L}{(1+r)2} + \dfrac{\ln r}{2}\right]} \right|^2 \tag{3.77}$$

Here, the pump intensity ratio at maximum reflectance $r = 1$. In addition, the probe transmittance is

$$T^+ = \left| \frac{\sinh\left(\dfrac{\ln r}{2}\right)\exp\left[\dfrac{(1-r)\,\gamma L}{(1+r)\,2}\right]}{\sinh\left[\dfrac{(1-r)\,\gamma L}{(1+r)\,2}+\dfrac{\ln r}{2}\right]} \right|^2 \tag{3.78}$$

Incidentally, as mentioned before, $\psi = \pi/2$ applies only when there is no externally-impressed electric field, and there is only carrier scattering, and where the carrier drift component is optimized via a moving diffraction grating as discussed in Section 3.3(8).

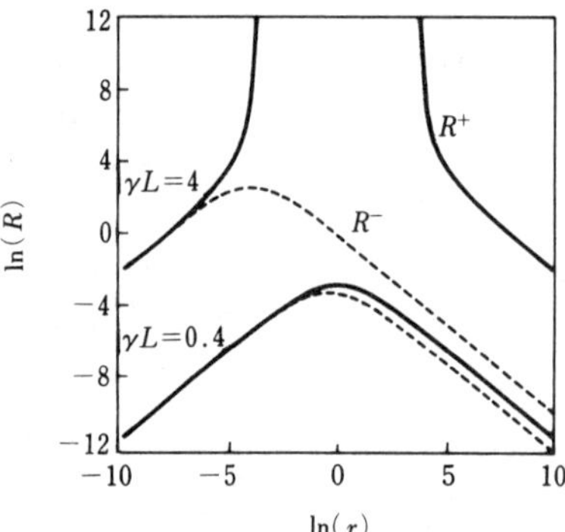

Figure 3.13 Reflectance R and pump intensity ratio r in an excited photorefractive medium. *(After Fischer and Weiss)*[23]

$r = I_b/I_f$, phase difference between the optical intensity distribution and refractive-index distribution $\psi = \pi/2$; the value of R^- (dashed curves) is obtained from Equation (3.74); R^+ (solid line) is obtained from Equation (3.77). γ: Complex coupling coefficient, L: length of medium

Figure 3.13 gives numerical values of the reflectance with positive and negative feedback in a photorefractive medium.[23] We see from this that the maximum reflectance is achieved with asymmetric pump intensity when there is negative feedback. Since the back pump wave is converted to phase-conjugate light in a transmitting medium, the stronger the back pump, the more effective it is in increasing R. On the other hand, when $r = 1$ for positive feedback (that is, the pump intensities are equal), the reflectance is a maximum and we may also get oscillation. In point of fact, there is no oscillation because of pump depletion.

(6) Effects of Absorption and Pump Depletion

An analytic solution allowing for both pump depletion and absorption within a medium in degenerate four-wave mixing can be found for light intensities in reflection[24] and transmission diffraction gratings.[25] An example of results for a reflection-type grating is shown in Figure 3.14. We see from this that, when the intensities of the probe and pump waves are of the same order, the pump intensities I_f and I_b are somewhat depleted in the medium when $R \leq 1$, and that the pump intensities cannot be neglected here.

Let us consider pump-wave depletion in degenerate four-wave mixing and find the analytic solution when absorption by the medium is neglected. If the

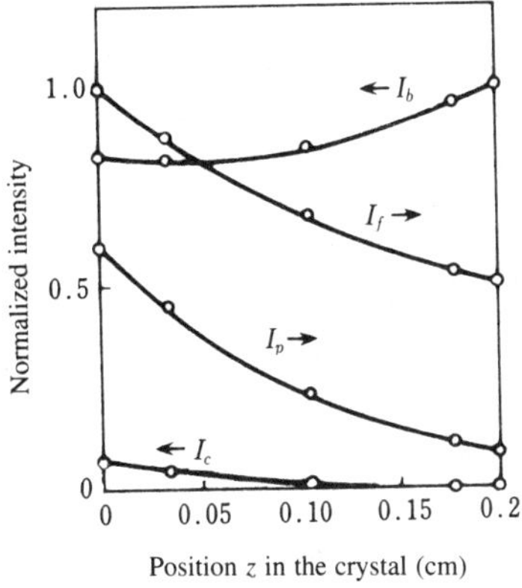

Figure 3.14 Intensities of the waves in a photorefractive medium when absorption and depletion are allowed for (degenerate four-wave mixing with counterpropagating waves, reflection-type grating). *(After Belić)*[24]

I_f and I_b, pump wave intensities; I_p, probe intensity; I_c, conjugate wave intensity. Arrows indicate propagation directions. Coupling coefficient $\varkappa = 6$ cm^{-1}; absorption coefficient $\alpha_0 = 3$ cm^{-1}.

arrangement is the same as that in Figure 3.1, we get coupled-mode Equation (3.70). If the probe intensity is I_p and the conjugate intensity is I_c, then the invariants are $(I_f + I_p)$, $(I_b + I_c)$, and $A_f A_b + A_p A_c = \xi$. The reflectance R can thus be obtained as follows.[26]

$$R = \frac{4\,|\,\xi\,|^2\,|\tanh(\mu L)\,|^2}{|\,\varDelta\tanh(\mu L) + (\varDelta^2 + 4\,|\,\xi\,|^2)^{1/2}\,|^2} \tag{3.79}$$

where $\varDelta = (I_b + I_c) - (I_f + I_p)$, $\mu = [\gamma(\varDelta^2 + 4|\xi|^2)^{1/2}]/2I_0$, γ is the complex coupling coefficient from Equation (3.71), I_0 is the total optical intensity, L is the thickness of the medium, and ξ is derived from the following equation.

$$[\,|\,\xi\,|^2 - I_f(0)\,I_b(L)\,]\,|\,\varDelta\tanh(\mu L) + (\varDelta^2 + 4\,|\,\xi\,|^2)^{1/2}\,|^2$$

$$+\,4\,|\,\xi\,|^2\,|\tanh(\mu L)\,|^2\,I_p(0)\,I_b(L)$$

$$+\,2\,|\,\xi\,|^2\,I_p(0)\,(\varDelta^2 + 4\,|\,\xi\,|^2)^{1/2}\{\tanh(\mu L) + [\tanh(\mu L)]^*\} = 0 \tag{3.80}$$

Figure 3.15 is a plot of the reflectance when pump depletion is taken into consideration.[26] The maximum value of the reflectance agrees with the value calculated from Equation (3.76) when the phase difference $\psi = 0$, and when there is no pump depletion. When pump depletion is taken into consideration, the reflectance shows damping and the coupling intensity increases. $R = 2.5$ in the illustration is exactly equivalent to the case where the total optical power of the back pump is converted to phase-conjugate light. In theory, the damping is produced because the phase of the light wave, which is the source of energy exchange between the light waves, is dependent on the light intensity.[27] In addition, when $|\gamma| L > 2$, the dependence of I_c on I_p becomes multi-valued and only one of the branches is the stable solution.

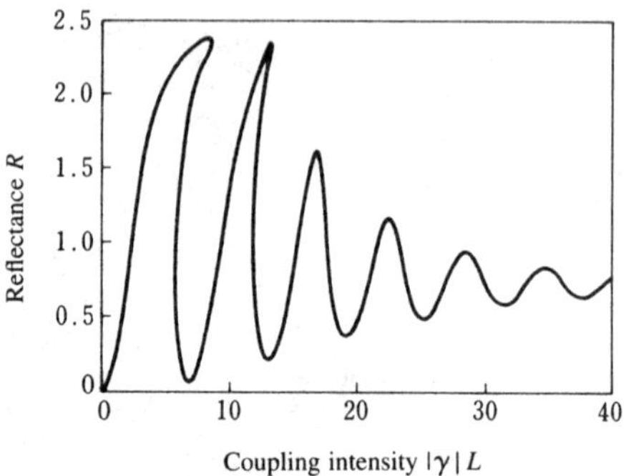

Coupling intensity $|\gamma|L$

Figure 3.15 Reflectance of a photorefractive medium when pump depletion is considered. *(After Cronin-Golomb, White, Fischer, and Yariv)*[26]

γ is the complex coupling coefficient from Equation (3.71), L is the medium thickness, $I_f(0) = I_b(L)$, $I_p(0) = 0.4I_b(L)$. Phase difference between optical intensity distribution and diffraction grating $\psi = 5°$.

(7) Transient Response of the Photorefractive Effect

The transient response of the photorefractive effect depends heavily on the efficiency of electric-field generation and the transport process because, after distribution of the electric field according to the illumination process, the spatial field is formed, changing the refractive index through the Pockels effect. If the transient response of the spatial electric field up to the time the steady-state value (3.67) is reached is represented by

$$E_{s2} = (I_2/I_0)\,E_m \exp(i\psi)\left\{1 - \exp\left[-\left(\frac{1}{\tau_g} + i\omega_g\right)t\right]\right\} \tag{3.81}$$

then we get the following equations for the time τ_g needed to establish the diffraction grating and the grating oscillation frequency ω_g.[28]

$$\tau_g = \frac{\tau_{di}\{[1 + (\tau_R/\tau_T)]^2 + (\tau_R/\tau_D)^2\}}{[1 + (\tau_{di}\tau_R/\tau_T\tau_I)][1 + (\tau_R/\tau_T)] + (\tau_R/\tau_D)^2(\tau_{di}/\tau_I)} \tag{3.82}$$

$$\omega_g = \frac{(\tau_R/\tau_T)[(\tau_{di}/\tau_I) - 1]}{\tau_{di}\{[1 + (\tau_R/\tau_T)]^2 + (\tau_R/\tau_D)^2\}} \tag{3.83}$$

where $\tau_{di} = \varepsilon_s\varepsilon_0/(I_0\alpha\eta_e\tau_R e\mu)$ is the dielectric relaxation time of carriers generated by illumination, I_0 is the light intensity, α is the coefficient of absorption, η_e is the quantum efficiency, and μ is the mobility. $\tau_R = 1/(\gamma_R N_A)$ is the recombination time, and γ_R is the recombination coefficient. $\tau_T = e/(\mu k_B T k_G^2)$ is the scattering time, and k_G is the wave number of the diffraction grating. $\tau_D = 1/(\mu E_0 k_G)$ is the drift time, and E_0 is the externally impressed electric field. $\tau_1 = 1/(sI_0 + \gamma_R n_c)$ is the reciprocal of the carrier formation and recombination rate, s is the excitation cross section, and n_c is the carrier density.

The diffraction grating setup time τ_g is almost determined by the dielectric relaxation time τ_{di} when the carrier recombination time τ_R is sufficiently short. Consequently, the response becomes faster to the extent that the incident light intensity I_0 increases. In addition, since the grating wave number k_G is contained in τ_T and τ_D, τ_G is also dependent on the grating interval Λ.

(8) Moving Diffraction Gratings

Four-wave mixing based on the photorefractive effect exerts a great influence on the reflectance through the phase difference ψ between the optical intensity distribution and the refractive-index distribution. When an external electric field E_0 is impressed on a medium, the drift component also increases and ψ increases from 0 to π, and the reflectance drops even though the maximum value E_q of the spatial electric field amplitude increases. Consequently, it is believed that a type of resonance effect can be used to increase the reflectance.

As shown in Figure 3.16, the response to the intensity distribution is that the carriers (electrons) are excited from the donor level to the conduction band. These carriers drift in the external electric field. At this time the interference fringes move at a certain constant speed by some mechanism; the speed of movement is coordinated with the carrier drift speed and the grating spacing. The movement of the fringes (i) can make use of piezoelements, where one wave causes mechanical vibration, and (ii) can be effected by detuning between the pump and probe. If this is done, the carriers accompany the movement of

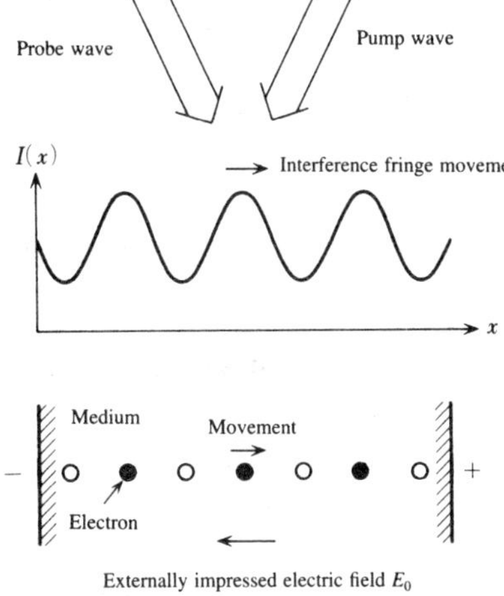

Figure 3.16 Moving diffraction grating in a photorefractive medium.

If an external electric field is impressed on a medium, electrons excited into the conduction band migrate spatially in response to the optical intensity distribution $I(x)$. If the optical interference migrates in synchrony with the electron movement, the optical energy moves efficiently with the carriers, and the reflectance rises.

the interference fringes, and are normally maximized. This result includes the phase effects, and a maximum spatial electric field is achieved. The $\psi = \pi/2$ component increases because the migration speed and spacing are optimized. The spatial electric field increases from mE_0 (where m is the spatial modulation) to $m(E_q/2)$ because of interference migration.[29] There is a factor-of-10 improvement when $E_0 = 10$ kV/cm with BSO. This type of diffraction grating is called a moving grating or running grating. Plans for improving the reflectance are being made in accordance with this.

The optimization of beam migration speed for grating spacing, externally-impressed electric fields and optical intensity is accomplished via two-photon coupling theory, and the results obtained have been quite good.[30,31]

(9) Other Characteristics

The theory of four-wave mixing based on the photorefractive effect has been analyzed with coupled-mode equations and other techniques, without delving into the origins of refractive-index changes and phase differences. The characteristics associated with detuning between pump and probe waves are the same as those for optical Kerr media:[32] the reflectance decreases dramatically with detuning.

Even when crystals are anisotropic, they can be normally dealt with by defining effective electro-optic coefficients. We must consider the tensor characteristics of the electro-optic coefficients in order to analyze polarization dependency. In this case, a different result is obtained for each crystalline symmetry. The species $BaTiO_3$[13] and BSO[33] have been dealt with. When optical activity is considered, the characteristics are numerically derived.[34]

3.4 Relationship Between Four-Wave Mixing and Other Phenomena

Even though degenerate four-wave mixing based on the optical Kerr effect and the refractive index uses the same terminology as four-wave mixing, the mechanisms are different. In addition, there is also a resemblance to holography. Consequently, the similarities and differences are shown in Table 3.2, and we will present a detailed explanation of them in this section.

(1) Relationship Between the Optical Kerr Effect and the Photorefractive Effect

First, we look at the differences in mechanism between the optical Kerr effect and the photorefractive effect in the generation of phase-conjugate light. The optical Kerr effect is a so-called $\chi^{(3)}$ process since the electron response of atomic vapors and anisotropic molecular liquids is used. The nonlinear coefficient of optical susceptibility here is based on the interaction of the oscillating dipole

Table 3.2 Relationships between four-wave mixing and holography.

	Four-wave mixing		Holography
	Optical Kerr effect	Photorefractive effect	
Mechanism	$\chi^{(3)}$ process based on electron response		
Recording and reconstruction		Simultaneous recording and reconstruction	Recording and reconstruction separate
		Recording and reconstruction frequencies may be different	Recording and reconstruction frequencies must be the same
Optical coupling intensity	Depends on optical intensity	Depends on electric field amplification	
Diffraction grating type	Spatial diffraction grating, temporal diffraction grating		Spatial diffraction grating only
	Necessary for four photons to exist simultaneously		Two photons exist simultaneously
Spatial phase shift between optical intensity distribution and refractive-index distribution	No	Yes	
Response speed	Comparatively fast	Generally slow	

induced by the electric field with the field itself. This interaction becomes strong as resonance occurs when the frequency of the incident light is near the energy-level transition frequency within the atoms and molecules. Therefore, degenerate four-wave mixing due to the optical Kerr effect differs in a fundamental way from holography based on spatial data. On the other hand, the photorefractive effect makes use of a diffraction phenomenon due to spatial refractive-index changes made by the interaction of four light waves (a hologram). That is, recording and reconstruction are not divided into two stages as in normal holography. Hologram data stemming from the interference of a pair of light waves can be reconstructed simultaneously by separate light waves. This behavior has led naturally to the development of real-time holography.

Since the refractive index varies in proportion to the optical intensity distribution, there is no spatial phase shift between the two for the optical Kerr effect. However, in the photorefractive effect, as explained in Section 3.3, there is a spatial phase shift between the distributions of optical intensity and refractive index to allow for the Poisson equations. A two-beam combination is possible because of this phase shift. In addition, data recording and reconstruction take place simultaneously because of this two-beam combination, thus making real-time holography possible.

Let us compare response speeds in this connection. With the optical Kerr effect, the response speed is fast since the change in index of refraction is mainly due to the electron response. However, with the photorefractive effect, since the refractive index changes (Pockels effect) after the spatial electric field is formed by the electric-charge distribution, the response speed strongly depends on the efficiency of charge generation and the transport process. The response is generally slow.

As expected from the mechanisms described above, the coupling coefficient with the optical Kerr effect depends on the electric field of both pump waves (optical intensity), in accordance with Equation (3.23). However, with the photorefractive effect, since the refractive index depends on the electric-field amplification ratio of each wave as shown by Equation (3.71), the coupling intensity is not dependent on the total optical intensity. As explained below, this difference causes a dramatic increase in the $\chi^{(3)}$ value equivalent to the photorefractive effect.

We use the results found in Sections 3.2 and 3.3, and study both differences quantitatively. We assume here degenerate four-wave mixing in an isotropic medium, and compare it to the complex amplitude of optical polarization $P_c^{(3)}$ in phase-conjugate light generation. From Equation (3.13) we get

$$P_c^{(3)} = 12\chi^{(3)} A_f A_b A_p{}^*$$

$$(3.84)$$

for the case of the optical Kerr medium, where $\chi^{(3)}$ is the third-order coefficient of nonlinear optical susceptibility and the A_j are the complex amplitudes. On the other hand, the change in the refractive index (the second term in Equation (3.62)) corresponds to the optically nonlinear term of Equation (3.16) for the photorefractive effect. If we use Equation (3.72) and Equation (C.1) of Appendix C, we can write

$$P = -2r_{\text{eff}} n^3 n_B E_{st} A_j$$

$$(3.85)$$

for the complex amplitude of the nonlinear polarization. Since the notation is the same as in Section 3.3, we omit it here. Because the optical interference term is valid in the photorefractive effect, Equation (3.68) can be used as the steady electric field E_{st}. Hence

$$P_c^{(3)} = -2r_{\text{eff}} n^3 n_B E_m \exp(i\psi)\,(1/I_0)\,(nc\varepsilon_0/2)\,A_f A_b A_p{}^*$$

$$(3.86)$$

is the polarization that contributes to phase-conjugate light generation. As mentioned before, polarization in photorefraction differs from the optical Kerr effect, and depends on the optical intensity ratio $(nc\varepsilon_0/2)A_f A_b /I_0$.

If we compare Equations (3.84) and (3.86) when the values of both optical nonlinear polarizations are equal, we get an equivalent nonlinear coefficient of susceptibility for the photorefractive index, that is,

$$\chi^{(3)} = \frac{cn^5\varepsilon_0}{12 I_0} r_{\text{eff}} E_q \left[\frac{E_0{}^2 + E_T{}^2}{E_0{}^2 + (E_q + E_T)^2} \right]^{1/2} \exp(i\psi)$$

$$(3.87)$$

where I_0 is the total optical intensity, r_{eff} is the electro-optic coefficient, E_q is the maximum spatial electric field, E_T is the scattering field, E_0 is the externally impressed dc field, and ψ is the phase difference between the optical intensity distribution and the distribution of the refractive index. Interaction is assumed to take place between the ordinary rays. The same relationship as in the above equation is found from the coupling equations for the probe and phase-conjugate waves, assuming that the coefficients of proportionality dA_p*/dz and A_c are equivalent in the optical Kerr effect and in the photorefractive effect.

Here BSO crystals are assumed, and if $E_q = 5 \times 10^5$ V/m, $n = 2.5$, $r_{\text{eff}} = 0.95 \times 10^{-12}$ m/V, and $I_0 = 10^4$ W/m^2 = 10 mW/mm^2, then $\chi^{(3)} = 1.0 \times 10^{-11}$ m^2/V^2 (7.4 $\times 10^{-4}$ esu), and a large coefficient of nonlinear susceptibility is obtained, equivalent to that in the photorefractive effect. Hence we see that the required optical energy is much smaller for the photorefractive effect than for the Kerr effect.

(2) Relationship Between Four-Wave Mixing and Holography

Holography is divided into a recording step and a reconstruction step, as shown in Figure 3.17. In the recording step, a reference wave (write wave) E_f and a data-laden object wave E_p simultaneously illuminate a hologram plate, and the interference pattern is recorded.[35] The transmittance is described by

$$T \propto (A_p + A_f)(A_p* + A_f*) = |A_p|^2 + |A_f|^2 + A_p A_f* + A_f A_p* \tag{3.88}$$

where A_j ($j = f, p, b, c$) are the complex amplitudes of the respective waves at the plane $z = 0$. In the reconstruction step, if separate reference waves (read waves) E_b and E_f are injected in exactly the reverse direction (so that the amplitudes A_p* and A_b are equal), the electric field E_c of the diffracted wave to the left of the hologram plate can be obtained from the following equation.

$$A_c = TA_b \propto (|A_p|^2 + |A_f|^2)A_f* + (A_f*)^2 A_p + |A_f|^2 A_p* \tag{3.89}$$

The first term on the right-hand side of the equation is proportional to the amplitude E_b of the reference wave in reconstruction; we are not interested in this term. The second term has a phase factor of exp $[-i (2k_f - k_p) \cdot r]$ and does not satisfy the phase-matching conditions. The third term is

$$A_c \propto |A_f|^2 A_p* = |A_f A_b| A_p* \tag{3.90}$$

It contains the phase factor exp $[i (k_f + k_b - k_p) \cdot r]$. Since $k_f + k_b = 0$ by hypothesis, this wave propagates in the direction opposite to the object wave E_p for $z < 0$ and is conjugate to it.

On the other hand, when the arrangement in Figure 3.1 is used for degenerate four-wave mixing, Equation (3.24b) yields

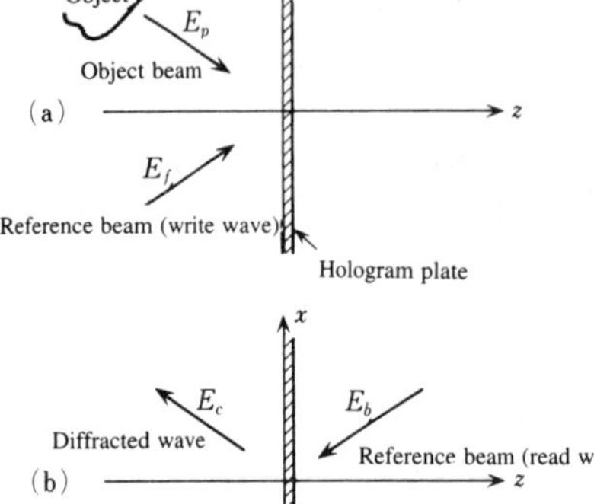

Figure 3.17 Data recording and reconstruction in holography.

(a) Recording step, (b) reconstruction step. E_b is incident in the direction opposite that of E_f.

$$A_c(z<0) = -i[(\varkappa^*/|\varkappa|)\tan(|\varkappa|L)]A_p{}^*(z<0)$$

(3.91)

for the complex amplitude of the phase-conjugate light. Here $\varkappa$ is the coupling coefficient. In particular, if $|\varkappa|L \ll 1$, we can rewrite Equation (3.91) as

$$A_c(z<0) \propto |A_fA_b|A_p{}^*(z<0)$$

(3.92)

Equations (3.90) and (3.92) agree, and we can see that holography and degenerate four-wave mixing have equivalent formulations. Since degenerate four-wave mixing does not separate recording and reconstruction as holography does, we can think of it as real-time holography.

Since holography is a two-step process, different frequencies can be used for recording and reconstruction in order to achieve magnification or reduction. In the recording step, however, the object and reference waves must have the same frequency. Otherwise, the interference pattern moves at a constant speed, the density of the photographic record is washed out, and the data cannot be reconstructed. On the other hand, recording and reconstruction take place simultaneously in four-wave mixing, although the moving grating introduces a Doppler shift when the frequencies are unequal, so that phase-conjugate light is produced. The pump and probe frequencies accordingly do not have to be equal in four-wave mixing; detuning merely lowers the reflectance.[11]

When four-wave mixing takes place in a nonlinear medium, characteristic phenomena of a tensorial nature result. As described in Section 3.2(1), the first and second terms in the nonlinear polarization Equation (3.13) represent the effects of the volume grating, and these terms are the same as for conventional holography. The third term represents the generation of phase-conjugate light by the temporal grating (steady state). This phenomenon is characteristic of four-wave mixing.[5]

4.

Special Theories of Four-Wave Mixing

Unlike Chapter 3, this chapter discusses special cases of four-wave mixing. In Section 4.1 resonant degenerate four-wave mixing theory is dealt with via the two-level atomic system model as a case where polarization cannot be represented as a series expansion. The quantum theory of degenerate four-wave mixing is discussed in Section 4.2, where light and media are dealt with quantitatively in a discussion of quantum noise. The theory of phase-conjugate light generation based on special mechanisms is covered in Section 4.3, and in Section 4.4 we discuss the transient response characteristics of outgoing waves in order to clarify their relationship to the response of media to injected light pulses.

4.1 Quasi-Classical Theory of Resonant Four-Wave Mixing

In the previous chapter, we discussed the case where polarization can be described by a power-series expansion of the electric field. This series expansion is appropriate for cases where the incident wave intensity is weak, and where the change in the number of carriers at each atomic or molecular energy level due to the incident waves is small. However,

(i) when a light frequency near the atomic or molecular resonance frequency is incident on a body that exhibits saturation absorption (including atomic vapor and organic dyes), or

(ii) when light is injected in the vicinity of a forward-biased semiconductor band gap many carriers are excited by resonance effects when the wave intensity is high. The number of carriers at the lower energy level dramatically decreases and the difference in the number of carriers at the upper and lower levels decreases. As a result, the rate of absorption shows a relative decrease, and saturation effects emerge. This phenomenon is called saturation absorption. The effects of saturation on dispersion are then described by the Kramers-Kronig relationship. In order to deal with polarization through the use of general methods in the first half of this section, we will use a two-level atomic system as the atomic and molecular model. We will treat polarization as an interaction between this system and the electric field (treated as a classical oscillator). This treatment corresponds to the theoretical extension of the optical Kerr effect of Section 3.2. This is very important in regard to applications to nonlinear polarization.

(1) Qualitative Explanation of Resonant Four-Wave Mixing

Here, also, the arrangement is the same as that shown in Figure 3.1, and the qualitative explanation is similar to that in Figure 3.2 (a) and (b). However, the method for setting up the diffraction grating is different. In the case of the transmission grating of Figure 3.2(a), E_f and E_p interfere, creating a spatial distribution of light intensity. Since absorption and dispersion produced by the two-level atomic optical absorption are dependent on optical intensity, a spatial distribution of the complex refractive index is formed in the medium. Here, E_b is diffracted by the grating and a new light wave (E_c, phase-conjugate to E_p) emerges in the direction opposite that of E_p due to the phase-matching conditions. The reflection grating of Figure 3.2(b) can also be explained by phase-conjugate light generation by the spatial distribution of the complex refractive index. In the case of the optical Kerr effect, the real part of the refractive index makes a contribution, but this case differs in that it is the complex refractive index (that is, absorption) that is effective.

(2) Derivation of Polarization from Classical Theory

We will describe material systems in terms of quantum theory and optical

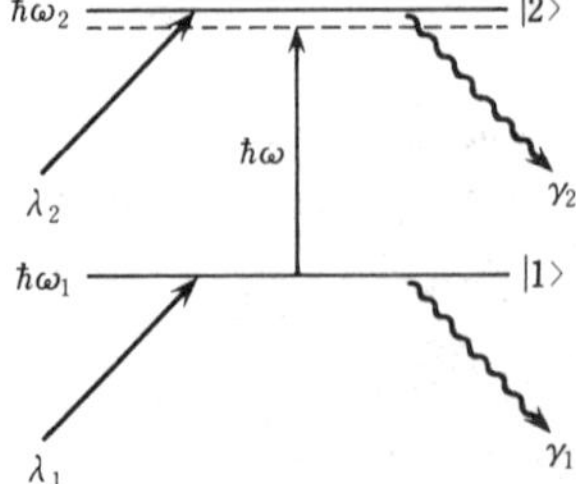

Figure 4.1 Atomic energy levels in the two-level model.

$\omega_R = \omega_2 - \omega_1$: atomic resonance frequency;
ω : incident light frequency;
λ_n: pumping rate for each level;
γ_n: relaxation rate for each level

systems in terms of classical theory in order to explain these points. As shown in Figure 4.1, we assume a two-level atomic system and look for the nonlinear polarization as a density matrix solution. The nonlinear coefficient of optical susceptibility in this case is based on the interaction between the excited electric dipole moments and the electric field. The diagonal components of the density matrix are the carrier-existence probabilities at each level, while the off-diagonal components correspond to the atomic coherence between the two levels. Although the off-diagonal component is zero when there are no light waves, it becomes non-zero when there are light waves, and contributes to nonlinear polarization.

The total polarization can be calculated from the following equation.[1]

$$P(\mathbf{r}, t) = \int_{-\infty}^{\infty} \rho_{12}(\mathbf{r}, \mathbf{v}, t)\, \mu_{21}\, dv + \text{c.c.} \tag{4.1}$$

where ρ_{12} is the off-diagonal component of the density matrix and μ_{21} [Cm] is the electric dipole moment. If we ignore the atomic translational motion, the density-matrix equation of motion can be expressed as follows:

$$i\hbar \frac{\partial \rho}{\partial t} = [H_0, \rho] + [V, \rho] + i\hbar\Lambda - \frac{i\hbar}{2}\{\Gamma, \rho\} \tag{4.2}$$

where $[\ ,\ ]$ is the commutator, $\{\ ,\ \}$ is the anticommutator, and H_0 is the unperturbed two-level atomic Hamiltonian satisfying the following equation.

$$H_0|n\rangle = \hbar\omega_n|n\rangle \tag{4.3}$$

Here $|n\rangle$ is the photon number state; ω_n is a fixed energy value, where the subscript $n = 1$ (2) denotes the ground (excited) state; $\hbar \equiv h/2\pi$, where h is Planck's constant; the lower atomic frequency is $\omega_R \equiv \omega_2 - \omega_1$; V is the interaction energy between the atoms and the light wave; Λ is the pumping rate to each level; and Γ is the relaxation rate of each level. Using E as the total electric field, the parameters are obtained from the following equations.

$$V = \begin{bmatrix} 0 & V_{12} \\ V_{21} & 0 \end{bmatrix}, \qquad \Lambda = \begin{bmatrix} \lambda_1 & 0 \\ 0 & \lambda_2 \end{bmatrix}, \qquad \Gamma = \begin{bmatrix} \gamma_1 & 0 \\ 0 & \gamma_2 \end{bmatrix} \tag{4.4}$$

where

$$V_{ij} = -\mu_{ij}E(\mathbf{r}, t) \tag{4.5}$$

Next, we look at the density matrix for the optical configuration of Figure 3.1. For simplicity we take the total electric field as

$$E(\mathbf{r}, t) = (1/2)\sum_j A_j(\mathbf{r})\exp[i(\omega t - \mathbf{k}_j \cdot \mathbf{r})] + \text{c.c.} \tag{4.6}$$

with equivalent polarization for all light waves, where the A_j [V/m] ($j = f, b, p, c$) are the complex amplitudes of the waves, and k_j [m^{-1}] is the wave vector. The optical intensities of the two pump waves here are assumed to be sufficiently great when compared to the intensities of the probe and phase-conjugate waves. The probe and the phase-conjugate waves in this case can be dealt with as a perturbation term for the pump waves. Consequently, the interaction energy and the density matrix become

$$V = V^{(0)} + qV^{(1)}, \qquad \rho = \rho^{(0)} + q\rho^{(1)} \tag{4.7a}$$

$$V^{(0)} = -(1/2)\,\mu\,E_0\exp(i\omega t) + \text{c.c.} \tag{4.7b}$$

$$V^{(1)} = -(1/2)\,\mu\,E_1\exp(i\omega t) + \text{c.c.} \tag{4.7c}$$

$$E_0 = A_f\exp(-i\boldsymbol{k}_f\cdot\boldsymbol{r}) + A_b\exp(-i\boldsymbol{k}_b\cdot\boldsymbol{r}),$$

$$E_1 = A_p\exp(-i\boldsymbol{k}_p\cdot\boldsymbol{r}) + A_c\exp(-i\boldsymbol{k}_c\cdot\boldsymbol{r}) \tag{4.7d}$$

where E_0 is the sum of the two pump waves, E_1 is the sum of the probe and the phase-conjugate waves, and q represents the degree of translation, which we will assume to have a value of 1. If Equations (4.4) – (4.7) are substituted in Equation (4.2), and if q is adjusted to be of the same order, we obtain the zero- and first-degree terms as follows:

$$i\hbar\frac{\partial\rho^{(0)}}{\partial t} = [H_0,\,\rho^{(0)}] + [V^{(0)},\,\rho^{(0)}] - \frac{i\hbar}{2}\{\Gamma,\,\rho^{(0)}\} + i\hbar\Lambda \tag{4.8}$$

$$i\hbar\frac{\partial\rho^{(1)}}{\partial t} = [H_0,\,\rho^{(1)}] + [V^{(0)},\,\rho^{(1)}] + [V^{(1)},\,\rho^{(0)}] - \frac{i\hbar}{2}\{\Gamma,\,\rho^{(1)}\} \tag{4.9}$$

The steady-state solution of zero-degree Equation (4.8) can be solved precisely if a rotating-wave approximation is used.[2] (When the incident-wave frequency and the atomic resonance frequency are the same, only these terms are selected for approximation in imaginary space because they each apply only to waves propagating in the same direction.) Moreover, the first-degree Equation (4.9) makes use of the zero-degree solution, and by the same token, can be precisely derived by using the same rotating-wave approximation. These results are applied to Equation (4.1), and the optical polarization can be obtained from[3]

$$P = \varepsilon_0\chi_0(E_0 + E_1) - \frac{\varepsilon_0\chi_0(n_0c\varepsilon_0/2)\,E_0(E_0E_1{}^* + E_0{}^*E_1)}{I_{sat}^0\{1 + \delta^2 + [(n_0c\varepsilon_0/2)|E_0|^2/I_{sat}^0]\}} + \text{c.c.} \tag{4.10}$$

where

$$I_{sat}^0 = (n_0c\varepsilon_0/2)\,[\hbar^2/(\mu^2 T_1 T_2)] \tag{4.11}$$

$$\delta = (\omega - \omega_R)\,T_2 \tag{4.12}$$

$$\chi_0 = \frac{-(\alpha_0 n_0{}^2/2)}{k}\,\frac{(i + \delta)}{\{1 + \delta^2 + [(n_0c\varepsilon_0/2)|E_0|^2/I_{sat}^0]\}} \tag{4.13}$$

$$\alpha_0 = (\omega/n_0c)\,\delta N\,\mu^2\,T_2/\hbar\varepsilon_0 \tag{4.14}$$

Here I_{sat}^0 [W/m^2] is the saturation intensity at the center frequency of the absorption line, δ is the normalized detuning frequency of the incident-wave and atomic-resonance frequencies, χ_0 is the optical nonlinearity for electric field E_0, α_0 [m^{-1}] is the small-signal saturation coefficient, T_1 ($=\gamma_1^{-1} = \gamma_2^{-1}$) is the energy (vertical) relaxation time, T_2 ($\approx \gamma_p^{-1}$) is the phase (horizontal) relaxation time, γ_p is the phase-relaxation constant, $\delta N = \lambda_1/\gamma_1 - \lambda_2/\gamma_2$ [m^{-3}] is the population difference between the upper and lower levels at equilibrium, and n_0 is the refractive index at low intensities.

In Equation (4.10), $E_0(E_0 E_1{}^* + E_0{}^* E_1)$ corresponds to the third-order nonlinear polarization and $A_f A_b A_p{}^*$ describes the phase-conjugate wave. The real and imaginary parts of the generating source correspond to the existence of a diffraction grating having a complex index of refraction (dispersion and absorption). The polarization becomes large at resonance, that is, when the frequency of the incident wave is near the two-level atomic-resonance frequency.

(3) Reflectance in Resonant Four-Wave Mixing

We substitute the nonlinear polarization derived from (4.10) and Equation (4.6) in the wave equation, and apply the phase-matching condition for four waves:

$$\boldsymbol{k}_f + \boldsymbol{k}_b - \boldsymbol{k}_p - \boldsymbol{k}_c = 0 \tag{4.15}$$

In addition, we assume that there is no pump depletion. If we use an adiabatic approximation, we obtain the following coupled-mode equations:[3,4]

$$\frac{dA_p{}^*}{dz} = iKA_c - \frac{\alpha^*}{2}A_p{}^* \tag{4.16a}$$

$$\frac{dA_c}{dz} = iK^*A_p{}^* + \frac{\alpha}{2}A_c \tag{4.16b}$$

where α [m^{-1}] is the wave-intensity absorption coefficient when saturation and polarization are considered and K [m^{-1}] is the coupling coefficient. These are defined by the following equations:

$$\alpha = \alpha_0 \frac{(1-i\delta)\left[1+(I_f+I_b)/I_{sat}\right]}{(1+\delta^2)\left\{\left[1+(I_f+I_b)/I_{sat}\right]^2-(4I_fI_b)/I_{sat}^2\right\}^{3/2}} \equiv \alpha_R - i\alpha_I \tag{4.17}$$

$$K^* = \alpha_0 \frac{(i+\delta)\left[(I_fI_b)/I_{sat}^2\right]^{1/2}}{(1+\delta^2)\left\{\left[1+(I_f+I_b)/I_{sat}\right]^2-(4I_fI_b)/I_{sat}^2\right\}^{3/2}} \tag{4.18}$$

$$I_{sat} \equiv I_{sat}^0 (1+\delta^2) \tag{4.19}$$

where I_{sat} [W/m^2] is the frequency-dependent saturation intensity and I_j [W/m^2] ($j = f,b$) is the pump intensity. Equations (4.16a and b) agree in form with absorption Equations (3.31a and b). However, the saturation effects due to resonance are reflected by the coefficients α and K in Equations (4.16a and b), which are also described as atomic level parameters. The α_I in Equation (4.17) corresponds to the dielectric constant ε_I, which in turn depends on the wave intensity.

These are related by the expression $\alpha_l = \varepsilon_l\,\omega/n_0 c$.

If Equations (4.16a and b) are solved with boundary conditions $A_p(z = 0) \neq 0$ and $A_c\,(z = L) = 0$, the probe and phase-conjugate fields can be obtained by the following equations:

$$A_p{}^*(L) = \frac{\gamma A_p{}^*(0)\exp(-i\alpha_l L/2)}{\gamma\cos(\gamma L) + (\alpha_R/2)\sin(\gamma L)} \qquad (4.20a)$$

$$A_c(0) = \frac{-iK^* A_p{}^*(0)\tan(\gamma L)}{\gamma + (\alpha_R/2)\tan(\gamma L)} \qquad (4.20b)$$

where $\gamma = [|K|^2 - (\alpha_R/2^2)]^{1/2}$. Here the output conjugate wave has the same frequency as the pump and probe waves. This is called resonant degenerate four-wave mixing. If we use the above equations, we can derive the reflectance R for phase-conjugate light:[3]

$$R = \left|\frac{K^*\tan(\gamma L)}{\gamma + (\alpha_R/2)\tan(\gamma L)}\right|^2 \qquad (4.21)$$

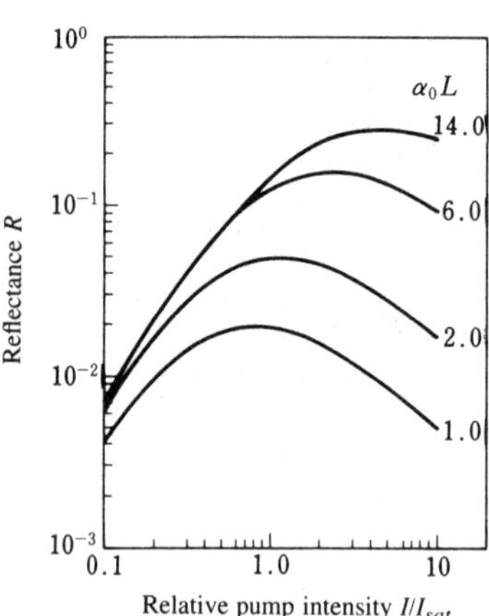

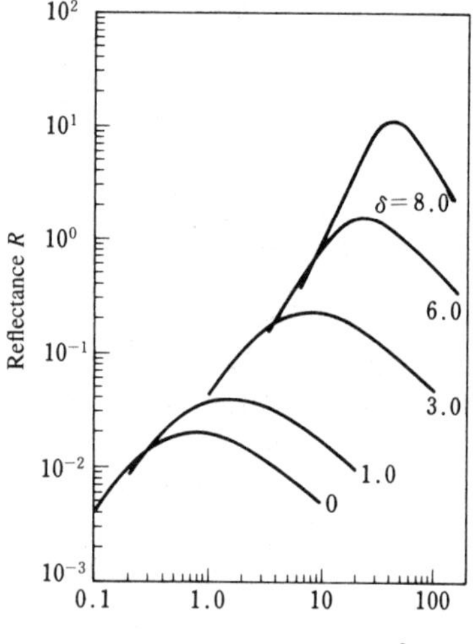

Figure 4.2 Phase-conjugate reflectance versus pump intensity in resonant absorption. (*After Abrams and Lind*)[3]

Normalized detuning frequency $\delta \equiv (\omega - \omega_R)T_2 = 0$;
ω : optical frequency;
ω_R : atomic resonance frequency;
     pump wave intensity: $I = I_f = I_b$;
I_{sat} : frequency-dependent saturation intensity;
α_0 : small-signal coefficient of absorption;
L : length of medium

Figure 4.3 Reflectance and normalized frequency shift δ in an absorbing medium. (*After Abrams and Lind*)[3]

Normalized detuning frequency $\delta \equiv (\omega - \omega_R)T_2$;
ω : optical frequency;
ω_R : atomic resonance frequency;
T_2 : phase relaxation time;
     pump wave intensity: $I = I_f = I_b$;
I_{sat}^0 : saturation light intensity at the center absorption frequency;
     $\beta L = 1.0$, $\beta = \alpha_0/(1 + \delta^2)$;
α_0 : small-signal intensity absorption coefficient;
L : length of medium

Although the reflectance shown in Equation (4.21) has the same form as Equation (3.35), an important feature of Equation (4.21) is that the absorption coefficient is expressed as a material parametric function, and is dependent on the wave intensity.

Figure 4.2 shows how the reflectance R depends on pump intensity at resonance.[3] The pump intensity is $I = I_f = I_b$ and the normalized (atomic) detuning frequency is $\delta = 0$ in Equation (4.1). The horizontal axis is normalized to the saturation intensity I_{sat}. We can see from the figure that if the pump intensity is near saturation, the reflectance is saturated, falling off at higher intensity. When $I > I_{sat}$, R decreases because excited carriers are overly increased due to resonance effects and the number of lower-level carriers to be excited is dramatically decreased. In addition, an increase in R due to an increase in the product $\alpha_0 L$ occurs because of an increase in the number of absorbed carriers which accompanies the increase in $\alpha_0 L$.

Figure 4.3 shows the relationship between reflectance in an absorbing medium and pump intensity. In particular, it shows the normalized detuning frequency δ as a parameter. The horizontal axis is normalized to saturation intensity I_{sat}^0 at the center of absorption. As the incident wave moves away from the atomic resonance frequency ω_R, the maximum reflectance increases. In addition, there is an increase in the pump intensity required for maximum reflectance. This reflects the fact that carrier saturation is less likely to occur when the incident wave is slightly detuned from ω_R.

From the above results, the plan for increasing the reflectance in a saturable medium is as follows:
(i) use waves having a frequency near the medium's resonant frequency;
(ii) make the small-signal intensity absorption coefficient α_0 large;
(iii) saturation can be attained with low optical power in a medium with a small I_{sat}^0, and is easy to achieve experimentally.

(4) Effects of Absorption and Pump Depletion

In the previous section we obtained an analytic solution for which it was assumed that the intensities of the two pumps were strong when compared to the probe and phase-conjugate waves. In addition, we neglected pump depletion. These assumptions lose relevance if the reflectance becomes large — especially if there is amplification — since the pump energy gradually migrates to the conjugate wave. The characteristics when both pump absorption and depletion are considered are derived by using iterative techniques (repetition of amplification of four combined waves until the boundary conditions are satisfied).

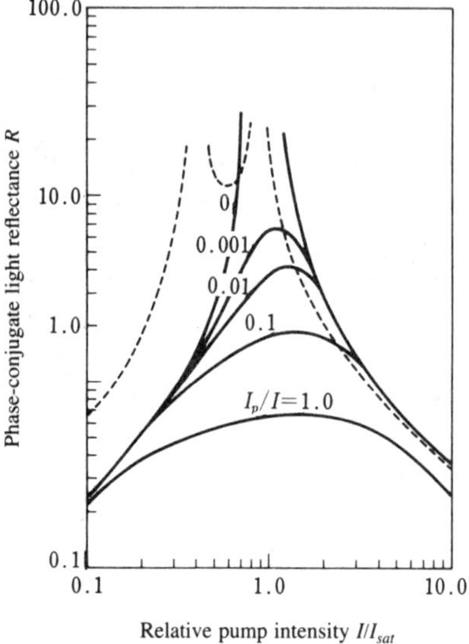

Figure 4.4 Pump depletion for phase-conjugate light reflectance during resonant degenerate four-wave mixing. *(After Brown)*[5]

Solid curve: with pump depletion
Dashed curve: constant pump intensity
Pump wave intensity: $I_f = I_b \equiv I$
Probe wave intensity: I_p
$\delta = (\omega - \omega_R)T_2 = 15$; ω: incident light frequency; ω_R: atomic resonance frequency; T_2: phase relaxation time

Figure 4.4 plots the reflectance for phase-conjugate waves versus probe intensity. The horizontal axis is normalized to the pump intensity.[5] Here $\delta = (\omega - \omega_R)T_2$, the normalized detuning between the incident wave and the atomic resonance frequency, is equal to 15; T_2 is the phase (horizontal) relaxation time. When the reflectance is 1, the effects of pump depletion can be neglected if the probe intensity is less than approximately 1/10 the pump intensity. However, if the probe intensity is at the same level as the pump intensity, the effects of pump depletion become striking.

(5) Nearly-Degenerate Four-Wave Mixing

Four-wave mixing with a small frequency difference between the probe and pump waves is called nearly-degenerate four-wave mixing, a model of which is shown in Figure 4.5. The nonlinear medium is modeled as a two-level atomic system whose resonant frequency is ω_R. The two pumps propagating from opposite directions (E_f and E_b) have a frequency of $\omega_0 = \omega_R + \Delta$, and the probe E_p has a frequency of $\omega_p = \omega_0 + \Omega$, where Δ is the shift in pump frequency from the resonant frequency and Ω is the shift in probe frequency from the pump frequency.

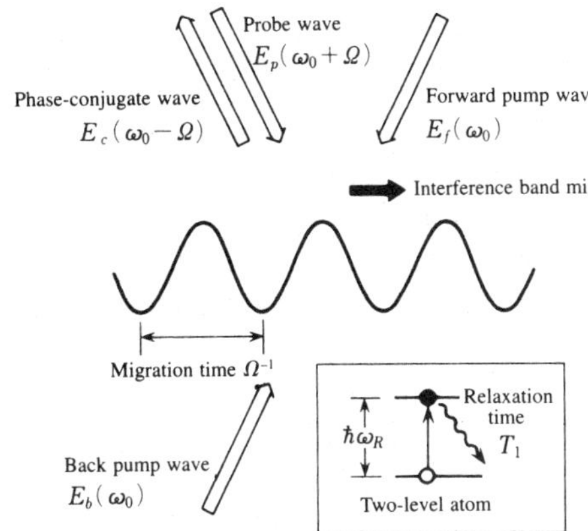

Figure 4.5 Interference fringe migration time (Ω^{-1}) for nearly-degenerate four-wave mixing in a resonant medium; vertical relaxation time (T_1) of the medium; relationship governing the reflectance.

Pump wave frequency: $\omega_0 = \omega_R + \Delta$
Probe wave frequency: $\omega_p = \omega_0 + \Omega$
ω_R: resonant frequency of the medium

When there is no shift, the excitation condition relaxes with a finite lifetime (vertical relaxation time) T_1, although a steady diffraction grating is formed in the nonlinear medium. On the other hand, if there is a shift, the interference fringes made by the forward pump E_f and the probe E_p (the diffraction grating) advance at a constant speed in the vector direction $\mathbf{k}_p - \mathbf{k}_f$. Since the back pump E_b is diffracted by the receding diffraction grating, phase-conjugate light E_c of frequency $\omega_0 - \Omega$ is produced due to the Doppler effect. The moving diffraction grating moves through one spatial period in time Ω^{-1}. Consequently, when T_1 is shorter than Ω^{-1}, the carriers (electrons) in the atoms effectively jump between the excited and ground states before the diffraction grating moves, thus increasing the reflectance. However, if Ω^{-1} is shorter than T_1, electron scattering occurs due to the motion of the diffraction grating, the spatial modulation rate of the diffraction grating becomes lower, and the reflectance decreases. If we measure the relationship between the reflectance and the probe wave shift (Ω), the inverse ($1/T_1$) of the atomic energy (vertical) relaxation time can be determined.[6] As a result, we can obtain spectroscopic information from the data relating the reflectance and the frequency. This is the nonlinear polarization technique.

The phase-conjugate light E_c that is generated via diffraction gratings formed by the back pump wave E_b and the probe wave E_p can be explained in a similar way.

We will now take up the characteristics when atomic translation motion is considered[7] and when the electric field is approximated not by plane waves but by Gaussian functions.[8]

4.2 Quantum Theory of Degenerate Four-Wave Mixing

In dealing with a nonclassical approach to squeezed photons, we will discuss classical theory under the idealization that the electric field is replaced by a boson operator. However, when studying the influences extending to the quantum effects of absorption, saturation and dispersion in a medium, material as well as optical systems must be treated in terms of quantum theory. In this section, we will develop the two-level atomic model with a nonlinear medium as an optical Kerr medium. Details of the derivation of the equations are explained in Appendix D(2).

(1) Quantum-Theoretic Derivation of Atomic Polarization

Take the same optical arrangement as in Figure 3.1, that is, two pump waves (E_f and E_b), probe wave E_p, and phase-conjugate wave E_c, all at the same frequency ω. These are described by the wave vectors k_j and the boson operators $\hat{a}_j$ ($j = f, b, p, c$ for the forward pump, back pump, probe and phase-conjugate waves, respectively); $\hat{a}^\dagger$ is the creation operator; and $\hat{a}$ is the annihilation operator. Assume that N two-level atoms are distributed in the medium, and that N_0 atoms are included in the volume element δV at location $\mathbf{r}$. We thus obtain a system of optical locations and atoms.

The principal equation for the density operator $\hat{\rho}$ uses a Markov approximation (a method in which the correlation function approximates the delta function when parametric fluctuations in certain systems are fast in comparison to other fluctuations), as represented by the following equation:[9]

$$\frac{\partial \hat{\rho}}{\partial t} = \frac{1}{i\hbar}[H_r, \hat{\rho}] + L_4[\hat{\rho}]$$

(4.22)

where the H_r are the Hamiltonians

$$H_r = H_1 + H_2 + H_3 \tag{4.23a}$$

$$H_1 = \sum_j \hbar\omega \hat{a}_j{}^\dagger \hat{a}_j \tag{4.23b}$$

$$H_2 = \sum_{m=1}^{N_0} (1/2)\hbar\omega_R \hat{\sigma}_{zm} \tag{4.23c}$$

$$H_3 = i\hbar g \sum_{m=1}^{N_0} (\hat{a}_r{}^\dagger \hat{\sigma}_m - \hat{a}_r \hat{\sigma}_m{}^\dagger) $$

$$L_4 = \sum_{m=1}^{N_0} \{(\gamma_1/2)([\hat{\sigma}_m \hat{\rho}, \hat{\sigma}_m{}^\dagger] + [\hat{\sigma}_m, \hat{\rho}\hat{\sigma}_m{}^\dagger]) \tag{4.23d}$$

$$+ (\gamma_p/4)([\hat{\sigma}_{zm}\hat{\rho}, \hat{\sigma}_{zm}] + [\hat{\sigma}_{zm}, \hat{\rho}\hat{\sigma}_{zm}])\} \tag{4.23e}$$

$$\hat{a}_r = \sum_j \hat{a}_j \exp(i\mathbf{k}_j \cdot \mathbf{r}) \tag{4.24}$$

The Hamiltonian H_1 is the sum of the energies of the four waves, H_2 is the energy of the atomic system in isolation, and H_3 is the interaction energy between waves and atoms. An electric dipole approximation and a rotating-wave approximation are used for H_3, where ω_R is the resonant frequency of the two-level atomic system and $\hat{\sigma}^\dagger$, $\hat{\sigma}$ and $\hat{\sigma}_z$ are the Pauli spin operators. The dipole coupling coefficient γ between wave positions and atoms is defined as $g = \mu E_u/2\hbar$. Here μ is the electric dipole moment, and E_u is the electric field strength referred to one photon. The photon energy is given by $\hbar\omega = n\varepsilon_0 E_u^2 V/2$, where V is the volume occupied by a photon. In addition, L_4 represents relaxation to a thermal bath due to natural atomic discharge and phase relaxation, $\gamma_1 = 1/T_1$ ($\gamma_{12} \approx \gamma_p = 1/T_2$) is the atomic vertical (horizontal) direction relaxation constant, γ_p is the phase relaxation constant, and T_1 (T_2) is the vertical (horizontal) relaxation time.

When the Fokker-Planck equation is written in order to find the excited atomic polarization at the optical location, the optical location for one atom is treated as a constant, provided the wavelength at the optical location is sufficiently long for a high-density medium approximation of the average inter-atomic spacing. The probability distribution function[10] is used, and the Fokker-Planck equation is obtained via Equation D.17 (Appendix). The relaxation of the atomic variable is assumed to be sufficiently faster than that of the optical position variable (adiabatic cancellation). The c for optical position $\hat{a}_j$ is assumed to be α_j. The two pump intensities (corresponding to α_f and α_b) are greater than the probe and the phase-conjugate wave intensities (α_p, α_c), and there is no pump depletion in the medium. If we keep only terms up to the first derivatives of the probe and phase-conjugate waves, we get the following equation for atomic polarization at position $\mathbf{r}$:[11]

$$v_r = -\frac{gN_0}{\gamma_{12}\Pi}(1-i\delta)(\xi+\zeta)\left[1-\frac{(nc\varepsilon_0/2)E_u^2(\xi^\dagger\zeta+\xi\zeta^\dagger)}{I_{sat}^0\Pi}\right]+\Gamma_r \tag{4.25}$$

where

$$\Pi = 1 + \delta^2 + \left[(nc\varepsilon_0/2)E_u^2|\xi|^2/I_{sat}^0\right] \tag{4.26a}$$

$$\xi = \alpha_f\exp(i\mathbf{k}_f\cdot\mathbf{r}) + \alpha_b\exp(i\mathbf{k}_b\cdot\mathbf{r}) \tag{4.26b}$$

$$\zeta = \alpha_p\exp(i\mathbf{k}_p\cdot\mathbf{r}) + \alpha_c\exp(i\mathbf{k}_c\cdot\mathbf{r}) \tag{4.26c}$$

The expression $igv_r n^2/kc$ in Equation (4.25) corresponds to the semi-classical polarization in Equation (4.10), which has a term corresponding to a third-order nonlinear polarization based on $(\xi+\zeta)(\xi^\dagger\zeta+\xi\zeta^\dagger)$. If the creation operator symbols are converted to complex conjugate notation, $\alpha_f\alpha_b\alpha_p{}^*$ becomes the term that contributes to phase-conjugate light generation. In addition, a new noise component Γ_r appears in the extended quantum treatment.

In the above equation, I_{sat}^0 is the pump-wave saturation intensity at the absorption nucleus, given by Equation (4.11) in accordance with classical theory;

$\delta = (\omega - \omega_R)T_2$ is the normalized detuning frequency; n is the index of refraction of the atomic system; c is the speed of light in vacuo; ε_0 is the permittivity; Γ_r is the non-zero noise component, where $<\Gamma_r(t)\Gamma_r(t')> = D_1\delta(t - t')$ and $<\Gamma_r^\dagger(t)\Gamma_r(t')> = D_2\delta(t - t')$ are the correlation coefficients; and D_1 and D_2 are parameters depending on I^0_{sat} and δ.

(2) Derivation of the Coupled-Mode Equations

The Heisenberg equation of motion for the probe and phase-conjugate waves makes use of the interaction Hamiltonian H_3:

$$\frac{d\hat{a}_j(t)}{dt} = \frac{1}{i\hbar} \int \frac{d^3r}{\delta v} [\hat{a}_j, H_r] = \int \frac{d^3r}{\delta v} g\hat{S} \exp(-i\mathbf{k}_j \cdot \mathbf{r}) \tag{4.27}$$

where $\hat{S}$ is the atomic dipole moment operator corresponding to the speed of light c at position $\mathbf{r}$ as in v_r. The equation for the amplitude at the optical position is:

$$\frac{da_j(t)}{dt} = \int \frac{d^3r}{\delta v} gv_r(t) \exp(-i\mathbf{k}_j \cdot \mathbf{r}) \tag{4.28}$$

The integral on the right side of the Equation (4.28) results in terms which satisfy the phase-conjugate matching condition $\mathbf{k}_f + \mathbf{k}_b = \mathbf{k}_p + \mathbf{k}_c$. If a phase mismatch of $\delta\mathbf{k}$ is assumed, phase matching functionally contributed to by the δ function is satisfied for an effective interaction length $L \gg (2\pi/|\delta\mathbf{k}|)$. We assume for simplicity that the intensities I of both pump waves are equal ($|\alpha_f|^2 = |\alpha_b|^2$). The time dependencies of the variables lead to Equations D.19 (Appendix). In order to solve for the boundary conditions in degenerate four-wave mixing, it is necessary to change the position dependencies. If we make the change of variable $z = ct$ ($-z = ct$) for the forward (back) wave, we get the following fundamental equations via quantum theory.[11]

$$\frac{da_p^\dagger(z)}{dz} = iKa_c(z) - \frac{\alpha^*}{2} a_p^\dagger(z) + G_p^\dagger(z) \tag{4.29a}$$

$$\frac{da_c(z)}{dz} = iK^*a_p^\dagger(z) + \frac{\alpha}{2} a_c(z) + G_c(z) \tag{4.29b}$$

where K is the coupling coefficient and α is the optical intensity absorption coefficient defined by the following equations:

$$K^* \equiv \alpha_0 \frac{(i+\delta)(I/I_{sat})}{(1+\delta^2)[1+(4I/I_{sat})]^{3/2}} \equiv \chi_R + i\chi_I \tag{4.30}$$

$$\alpha \equiv \alpha_0 \frac{(1-i\delta)[1+(2I/I_{sat})]}{(1+\delta^2)[1+(4I/I_{sat})]^{3/2}} \equiv \alpha_R - i\alpha_I \tag{4.31}$$

where $I_{sat} \equiv I^0_{sat}(1 + \delta^2)$ is the pump-wave saturation intensity depending on detuning δ, and $\alpha_0 = 2g^2N_0/c\gamma_{12}$ is the small-signal intensity absorption coefficient. This agrees with the classical theoretical expression in Equation (4.14). In addition, $G_j(z)$ in Equation (4.29) is an antisymmetric function that satisfies the following equations:

$$\langle G_c(z)\,G_p(z')\rangle = -\,\Psi^*\delta(z-z') \tag{4.32a}$$

$$\langle G_c{}^\dagger(z)\,G_p{}^\dagger(z')\rangle = -\,\Psi\delta(z-z') \tag{4.32b}$$

$$\langle G_c(z)\,G_c{}^\dagger(z')\rangle = \langle G_p(z)\,G_p{}^\dagger(z')\rangle = \Phi\delta(z-z')$$
$$\Psi \equiv \Psi_R + i\Psi_I \tag{4.32c}$$

$$\Psi_R \equiv \frac{-\alpha_0}{(1+\delta^2)^2 I_3}\left[(1-3\delta^2)I_1 + (1+\delta^2)^2 I_2\right] \tag{4.32d}$$

$$\Psi_I \equiv \frac{-\alpha_0}{(1+\delta^2)^2 I_3}(3\delta - \delta^3)I_1 \tag{4.32e}$$

$$\Phi \equiv \frac{\alpha_0}{(1+\delta^2)^2 I_3}\left[12(1+\delta^2)(I/I_{sat})^2 + (1+\delta^2)^2 I_2\right] \tag{4.32f}$$

$$I_1 \equiv (2I/I_{sat})\left[1+(I/I_{sat})\right] \tag{4.32g}$$

$$I_2 \equiv (1/2)I_3 - (1/2) - (5I/I_{sat}) - 15(I/I_{sat})^2 \tag{4.32h}$$

$$I_3 \equiv \left[1+(4I/I_{sat})\right]^{5/2} \tag{4.32i}$$

If we compare Equations (4.29a and b) with classical Equations (4.16a and b), the complex conjugate symbol is replaced by the creation operator symbol. Moreover, a term relating to the antisymmetric function $G_j(z)$ is added, and a new noise component is generated in the quantum-theoretic treatment. As shown in Appendix D(1), the generation of a squeezed state having one optical quantum effect can be explained if the complex amplitude is replaced by an operator, even if the absorption and noise terms are ignored in Equations (4.29a and b). However, that case is the ideal condition, and the influence of absorption, saturation and dispersion cannot be analyzed if the more precise Equations (4.29a and b) are not used.

(3) Optical Position in Consideration of Quantum Noise

Now, after change of the variables in the probability differential Equations (4.29a and b) to

$$\alpha_p'(z) = \alpha_p(z)\exp(-i\alpha_I z/2) \tag{4.33a}$$

$$\alpha_c'(z) = \alpha_c(z)\exp(i\alpha_I z/2) \tag{4.33b}$$

we get the following equations[11,12] if we solve for the probe and phase-conjugate waves using standard methods with boundary conditions $\alpha_p(z=0)$ and $\alpha_c(z=L)$ at both ends of the medium.

$$\alpha_p(L) = \bar{T}\exp(i\alpha_I L/2)\,\alpha_p(0) + \bar{R}\alpha_c{}^\dagger(L) + F_p \tag{4.34a}$$

$$a_c(0) = \bar{T} \exp(i\alpha_l L/2)\, a_c(L) + \bar{R} a_p{}^\dagger(0) + F_c \qquad (4.34b)$$

where

$$\bar{T} = \frac{\gamma}{\gamma \cos(\gamma L) + (\alpha/2)\sin(\gamma L)} \qquad (4.35a)$$

$$\bar{R} = \frac{-iK^* \sin(\gamma L)}{\gamma \cos(\gamma L) + (\alpha/2)\sin(\gamma L)} \qquad (4.35b)$$

$$F_c = \int_0^L [G_c(z)\, a(z) + G_p{}^\dagger(z)\, b(z)]\, dz \qquad (4.35c)$$

$$F_p = \int_0^L [G_c{}^\dagger(z)\, c(z) + G_p(z)\, d(z)]\, dz \qquad (4.35d)$$

$$a(z) = -\bar{T} f_2(L-z), \qquad b(z) = \bar{T} f_1(L-z) \qquad (4.35e,f)$$

$$c(z) = \bar{R} \exp(i\alpha_l L)\, f_2(L-z) + f_1(L-z) \qquad (4.35g)$$

$$d(z) = \bar{R} \exp(i\alpha_l L)\, f_1(L-z) + f_2(L-z) \qquad (4.35h)$$

$$f_1(z) = (-iK^*/\gamma)\sin(\gamma z) \qquad (4.35i)$$

$$f_2(z) = \cos(\gamma z) + (\alpha/2\gamma)\sin(\gamma z) \qquad (4.35j)$$

$$\gamma = [\,|K|^2 - (\alpha/2)^2\,]^{1/2} \qquad (4.35k)$$

Even if phase-conjugate light does not exist at $z = L$, it is very important in Equation (4.34) that L not be 0 because of quantum noise caused by zero-point wave motion. $\bar{T}$ and $\bar{R}$ in Equations (4.35a and b) are the transmittance and the reflectance, respectively.

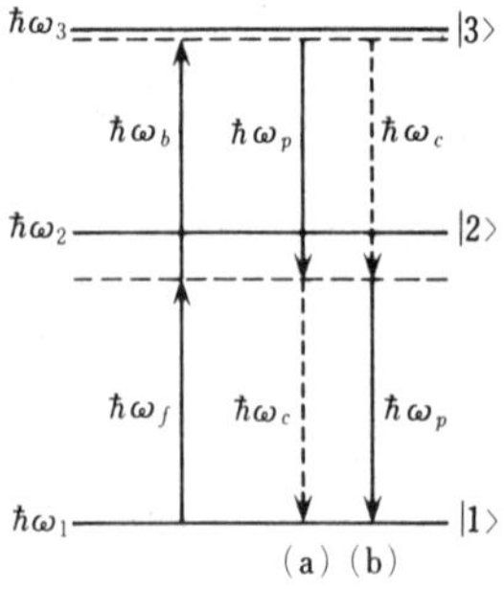

Figure 4.6 Resonant two-photon absorption in three-level atoms.

(a) Phase-conjugate light generation process due to pump and probe waves
(b) Probe wave generation process due to pump and phase-conjugate waves
$\hbar\omega_n$: Energy at each level ($n = 1 - 3$)
ω : Incident wave frequency
ω_{2R}: Migration frequency between levels 1> and 3>

4.3 Special Mechanisms in Four-Wave Mixing

We will get a grasp of four-wave mixing in a broad sense in this section, and will explain the theory of phase-conjugate light generation through resonant two-photon absorption, forward four-wave mixing, and photon echoes.

(1) Resonant Two-Photon Absorption

Assume that the optical arrangement is the same as that for counter-propagating degenerate four-wave mixing in Figure 3.1, and that the frequency of each wave is ω. Resonant two-photon absorption corresponds to the third term of Equation (3.13), where different coherence level spacings participate in the generation of phase-conjugate light. We will consider the three-level atoms in Figure 4.6 in order to explain this mechanism, wherein we assume a migration frequency of ω_{2R} between energy levels I1> and I3>. Here opposing pump waves form a coherence between levels I1> and I3>. This coherence excitation is at frequency 2ω, oscillates temporally and is spatially uniform. Resonance conditions for two-photon coherence assume a velocity vector v for each atom,[13] and are expressed by

$$2\omega - (k_f + k_b) \cdot v = \omega_{2R} \tag{4.36}$$

Since the two pump waves are propagating in opposite directions ($k_f + k_b = 0$), atoms moving in any direction can participate in resonant two-photon absorption. On the other hand, the probe wave induces the two-photon process; as a result, phase-conjugate light is produced by the process in Figure 4.6(a). Further, as shown in Figure 4.6(b), newly-generated phase-conjugate light is produced by the probe wave, making possible high-resolution polarization with no Doppler broadening. This is called Doppler-free polarization.

The polarization related to the two-photon absorption can be calculated from:

$$P(r, t) = \int_{-\infty}^{\infty} dv \left[\rho_{12}(r, v, t) \mu_{21} + \rho_{23}(r, v, t) \mu_{32} + \text{c.c.} \right] \tag{4.37}$$

where ρ_{ij} is the density matrix, μ_{ij} is the dipole moment, and the subscripts correspond to the energy levels. If we use the same technique as in Section 4.1(2), we obtain

$$P(r, t) \propto \frac{|\mu_{12}|^2 |\mu_{23}|^2}{\gamma_{13} + i(2\omega - \omega_{2R})} A_f A_b A_p{}^* \exp[i(\omega t + k_p \cdot r)] \tag{4.38}$$

as the polarization contributing to the production of phase-conjugate light. Here γ_{13} is the coherence relaxation index between levels I1> and I3> (the reciprocal of the horizontal relaxation time). From Equation (4.38), the spectrum of the generated phase-conjugate light has a Lorentzian shape with center frequency $\omega_{2R}/2$ and line width γ_{13}.

(2) Forward Four-Wave Mixing

Consider, as shown in Figure 4.7, pump and probe waves E_L and E_p of the same frequency ω propagating in the same direction as they are injected into a nonlinear medium (nonlinearity $\chi^{(3)}$ and thickness L). Here the excitation is proportional to $E_L E_p{}^*$ because of interaction between both waves in the medium, where E_L is already self-diffracting in the $2\boldsymbol{k}_L - \boldsymbol{k}_p$ direction and E_s is generated. We notice two forward-type incident beams (to be dealt with later). Since four photons participate in the generation of the diffracted light, it is classified as four-wave mixing. If we compare the above-mentioned forward four-wave mixing to the three-wave mixing of Section 5.1, we get the same limitations due to the phase-matching conditions. On the other hand, unlike three-wave mixing whose incident wave frequencies are ω and 2ω, the incident frequencies for the forward type are the same.

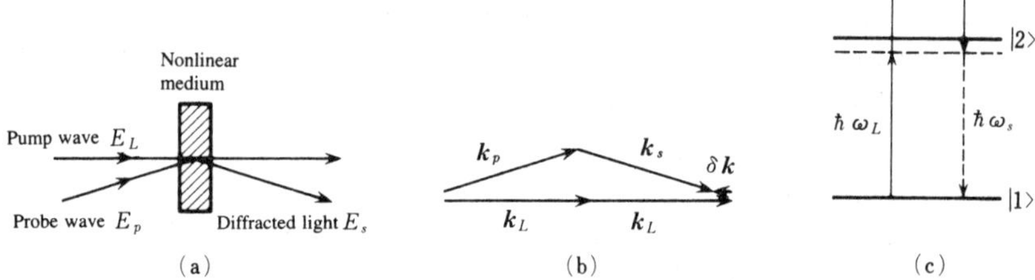

Figure 4.7 Principle of forward four-wave mixing.

(a) beam positions,
(b) phase-matching vectors,
(c) interaction between three-level atoms and waves
$\delta\boldsymbol{k} = \boldsymbol{k}_s + \boldsymbol{k}_p - 2\boldsymbol{k}_L$

In order to deal with four-wave mixing quantitatively, we represent the electric fields by

$$E_j(\boldsymbol{r},\ t) = (1/2)\,A_j(\boldsymbol{r})\exp[i(\omega_j t - \boldsymbol{k}_j\cdot\boldsymbol{r})] + \text{c.c.} \quad (j = L, p, s) \tag{4.39}$$

where A_j is the complex amplitude, $\boldsymbol{k}_j$ is the wave vector and $\boldsymbol{r}$ is the position vector. We obtain here

$$P_{NL} = (1/2)\,\varepsilon_0\chi^{(3)}\Big[3\sum_j |A_j|^2 A_j + 3\sum_{j\neq k} A_j \sum_k A_k^2$$

$$+ 6\sum_{j\neq k} A_j \sum_k |A_k|^2 + 6\sum_{j\neq k\neq l} A_j A_k A_l\Big] + \text{c.c.} \tag{4.40}$$

as the nonlinear polarization at frequency ω, where the subscripts represent any of L, p or s, and ε_0 is the permittivity in vacuo. $\chi^{(3)}$ is a constant independent of the component. We substitute this in the wave equation of motion (3.12) and use the adiabatic approximation. We select the terms for which the frequency and wave number agree on both sides to obtain the following coupled-mode equations:[14]

$$dA_L/dz + [(\alpha_0/2) - i\eta\,(|A_L|^2 + 2|A_p|^2 + 2|A_s|^2)]A_L$$
$$= 2i\eta A_L{}^* A_p A_s \exp(i|\delta k|z) \tag{4.41a}$$

$$dA_p/dz + [(\alpha_0/2) - i\eta\,(2|A_L|^2 + |A_p|^2 + 2|A_s|^2)]A_p$$
$$= i\eta A_L{}^2 A_s{}^* \exp(i|\delta k|z) \tag{4.41b}$$

$$dA_s/dz + [(\alpha_0/2) - i\eta\,(2|A_L|^2 + 2|A_p|^2 + |A_s|^2)]A_s$$
$$= i\eta A_L{}^2 A_p{}^* \exp(i|\delta k|z) \tag{4.41c}$$

$$\eta = 3\omega\chi^{(3)}/2nc \tag{4.42}$$

where $\delta k = k_p + k_s - 2k_L$ is the phase mismatch, $\alpha_0 = \sigma\mu_0 c/n$ is the small-signal optical intensity absorption coefficient, σ is the electric conductivity, n is the index of refraction of the medium, and c is the speed of light in vacuo. We see from Equation (4.41c) that the diffracted light is almost symmetric to the probe wave at a pump frequency ω. When the forward phase mismatch is very small, although the spatial phase of the diffracted light inside the surface perpendicular to the direction of propagation is the complex conjugate of the probe wave, the forward-type diffracted light is not phase-conjugate light in the strictest sense of the word because it has no time-reversal characteristics. When the third-order nonlinearity $\chi^{(3)}$ in Equation (4.42) is a real number, it is omitted since the solution[14] is similar to that in Section 3.2(4).

When the pump and probe waves are collinear, the phase mismatch δk can be seen to be approximately zero. Normally $I_p \ll I_L$, and the effects of the probe wave and diffracted light excitation can be ignored because they are much smaller than that of the pump wave. In addition, the first term can be neglected because it is very small in comparison to the second term for the pump wave. We then obtain the following coupled-mode equations from Equations (4.41):

$$dA_L/dz + [(\alpha_0/2) - i\eta|A_L|^2]A_L = 0 \tag{4.43a}$$

$$dA_p/dz + [(\alpha_0/2) - 2i\eta|A_L|^2]A_p = i\eta A_L{}^2 A_s{}^* \tag{4.43b}$$

$$dA_s/dz + [(\alpha_0/2) - 2i\eta|A_L|^2]A_s = i\eta A_L{}^2 A_p{}^* \tag{4.43c}$$

Equations (4.43) are the starting points for analysis of forward four-wave mixing. We will discuss below the case where $\chi^{(3)}$ is an imaginary number. The right-hand sides are the gain terms, and represent conversion of the two photons from pump wave to probe wave and diffracted light. The expression in brackets [ ] represents the effect of linear absorption and two-photon absorption. The imaginary part of $\chi^{(3)}$ contributes to two-photon absorption.

Since Equation (4.43a) contains only the pump wave, it can be treated independently of the other equations. We define the intensity as $I_j = (nc\varepsilon_0/2)|A_j|^2$, and obtain the following generalized Riccati differential equations:

$$dI_L/dz = (-\alpha_0 - \alpha_2 I_L)\,I_L \tag{4.44}$$

$$\alpha_2 = 6\omega\chi_I^{(3)}/\varepsilon_0(nc)^2 \tag{4.45}$$

where α_2 is the coefficient of two-photon absorption and $\chi_I^{(3)}$ is the imaginary part of the nonlinear coefficient of susceptibility. We obtain the following equation for the pump intensity by integrating Equation (4.44):

$$I_L(z) = I_L(0)\,\frac{\exp(-\alpha_0 z)}{1 + \alpha_2 I_L(0)\,\xi(z)} \tag{4.46}$$

where $\xi(z) \equiv [1 - \exp(-\alpha_0 z)]/\alpha_0$ is the effective interaction length.

If we solve Equation (4.43) using the experimentally-obtained optical intensity formula, we get the following equations:[14]

$$I_L(z) = \frac{I_L(0)\exp(-\alpha_0 z)}{F(z)} \tag{4.47a}$$

$$I_s(z) = \frac{1}{F}\,\hat{I}_p(z)\left[\frac{|\alpha|}{2}I_L(0)\,\xi(z)\right]^2 \tag{4.47b}$$

$$I_p(z) = \hat{I}_p(z)\left\{1 + \frac{1}{F}\left[\frac{|\alpha|}{2}I_L(0)\,\xi(z)\right]^2\right\} = \hat{I}_p(z) + I_s(z) \tag{4.47c}$$

$$F(z) \equiv 1 + \alpha_2 I_L(0)\,\xi(z) \tag{4.48}$$

$$\alpha = 6\omega\chi^{(3)}/\varepsilon_0(nc)^2 \tag{4.49}$$

where $\hat{I}_p(z)$, the probe wave solution when there is no gain in Equation (4.43), is given by

$$\hat{I}_p(z) = I_p(0)\,[I_L(z)/I_L(0)]^2\exp(-\alpha_0 z) \tag{4.50}$$

Equation (4.47b) expresses the competition between the diffracted light and gain, which depend on $|\chi^{(3)}|I_L(0)\xi(z)$, and two-photon absorption, which depends on F. Equation (4.47c) shows that pump power declines while probe power increases.

As shown in Figure 4.8, there is also a method for using three non-coplanar beams. If these beams are injected into a nonlinear medium, diffracted light is generated in the direction $k_s = k_{L1} + k_{L2} - k_p$.[15] There is a benefit even when thick test materials are used since the phase mismatch for 3-beam injection is much smaller than that for 2-beam injection.

(3) Photon Echoes

As shown in Figure 4.9, if a first pulse is injected into an optically non-linear medium of nonuniform width, and a second pulse is injected, after a time interval τ,

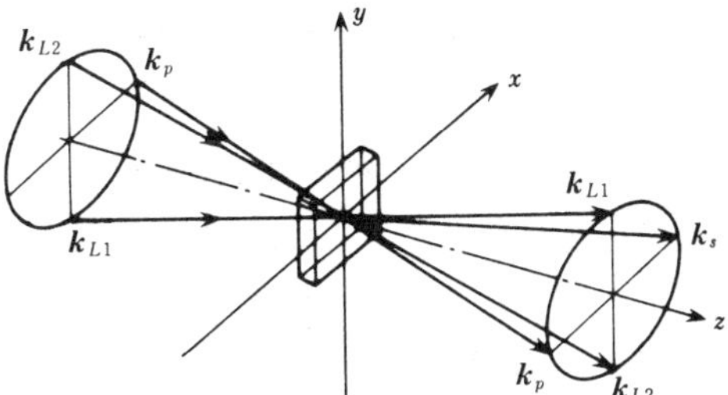

Figure 4.8 Forward four-wave mixing (not all beams are coplanar).

k_{L1}, k_{L2}: Pump wave vectors
k_p: Probe wave vector
k_s: Diffracted wave vector

then a photon echo pulse (the third pulse) is generated, again after a further time τ. If certain conditions are satisfied here the photon echo pulse becomes the conjugate of the first pulse. The equations explain the details of this.

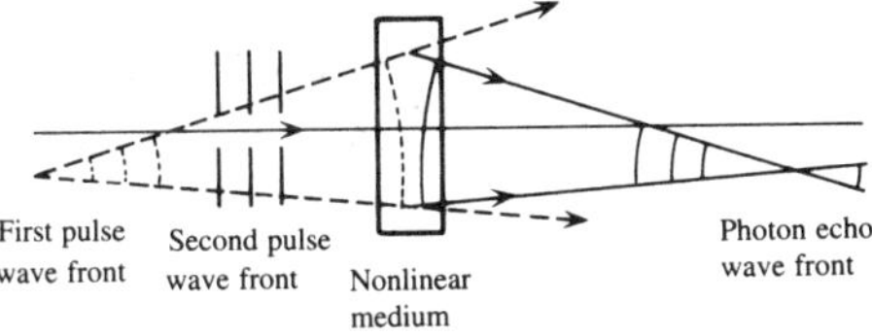

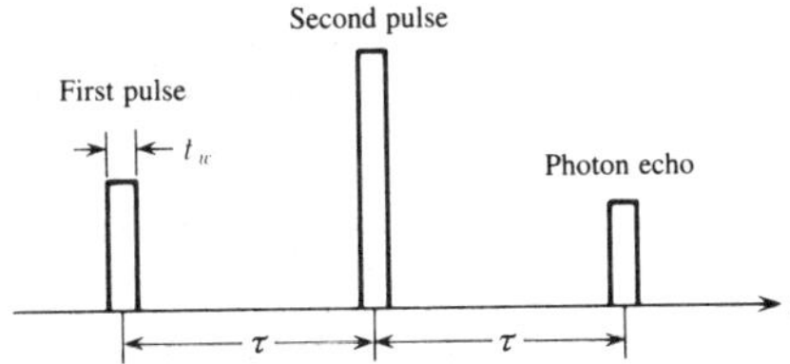

Figure 4.9 Example of phase-conjugate light production using photon echoes.

If the second pulse (π pulse) is injected a time τ after injection of the first pulse ($\pi/2$ pulse), a photon echo pulse is generated after further time τ when a steady state is reached.

The following equation expresses the electric field for each pulse:

$$E_j(\boldsymbol{r}, t) = (1/2)\, A_j(t) \exp[i(\omega t - \boldsymbol{k}_j \cdot \boldsymbol{r})] \exp[i\phi_j(\boldsymbol{r})] + \text{c.c.} \quad (j = 1\sim3)$$

$$(4.51)$$

where $A_j(t)$ is the complex amplitude and $\phi_j(r)$ is the spatial phase. If the first and second pulses are injected into a third-order medium having nonlinearity $\chi^{(3)}$, then the nonlinear polarization can be expressed by

$$P_{NL} \propto |E_1(r, t) + E_2(r, t - \tau)|^3 \tag{4.52}$$

Among the many polarization components, the frequency component ω produced by the time interval $t = 2\tau$ is

$$P_{NL} = (3/2)\,\varepsilon_0\chi^{(3)}A_1{}^*(t)\,A_2{}^2(t - \tau)\exp[i\omega(t - 2\tau)]$$

$$\cdot \exp[-i(2k_2 - k_1) \cdot r]\exp[i(2\phi_2 - \phi_1)] + \text{c.c.} \tag{4.53}$$

A photon echo pulse is generated after a time interval τ following the second pulse, due to this component. From Equation (4.53), this process can also be viewed as a type of four-wave mixing since there are one first pulse and two second pulses involved in the generation of an echo. When the spherical surface waves of the light pulses have curvature radius R_j, the wave fronts are related by the following equation:

$$1/R_3 = 2/R_2 - 1/R_1 \tag{4.54}$$

If the second pulse is a plane wave $(R_2 = \infty,\ \phi_2 = 0)$, then $\phi_3 = -\phi_1$, and the echo pulse becomes the phase conjugate of the first pulse.[16]

Incidentally, the pulse area becomes a problem when the photon echo component is optimized. If the pulse is square, and the pulse width is t_w, then the pulse area can be estimated by

$$\theta_j = 2t_w|\mu_{12}||A_j|/\hbar \tag{4.55}$$

where μ_{12} is the dipole moment between the two levels.[17] The pulses are called "$\pi/2$ pulses" or "π pulses," depending on the value of θ. The echo component is maximized when the first pulse has magnitude $\pi/2$ and the second pulse has magnitude π, even if the photon echo is generated using an arbitrary pulse magnitude.

There is also a case where the echo pulse is the first or second pulse's phase conjugate if the three pulses are injected in phase at different times.[18] In addition, there is a close relationship between photon echoes and normal four-wave mixing,[19] which we will not cover.

4.4 Transient Response Characteristics of Four-Wave Mixing

We will discuss first in this section how the time response of radiated light depends on the ratio between the pulse width of the incident light and the response time of the nonlinear medium. Then we will look at phase-conjugate light transient response characteristics for pulsed probe incidence when continuous pump waves are radiated into a nonlinear medium, using the transfer function. The explanation will be divided into optical Kerr media and photorefractive media. We will also touch upon their relationships.

(1) The Relationship Between Pulse Width and Medium Response Time
a. Introduction

When we consider the interaction between light and matter, unless we immediately grasp the relationship between incident light pulse width and substance response time (for example, atomic and molecular level lifetimes and phase relaxation time), there is always a risk of unexpected failure when measured values are compared with theoretical values.

Since adequate substance-system response is obtained if the incident light is a continuous wave or if the optical pulse width is sufficiently longer than the substance's response time, polarization due to the nonlinear interaction between the light waves and the substance attains a steady value. Consequently, the value obtained in this case can be used as is. We will try to apply this idea to degenerate four-wave mixing. If the substance's response time is sufficiently shorter than the probe wave's incident pulse width, and if we let the $A_j(t)$ denote the complex amplitudes of the waves, then the resulting polarization is given by

$$P_{NL} \propto \chi^{(3)} A_f(t) A_b(t) A_p{}^*(t) \tag{4.56}$$

and the pulse width of the generated phase-conjugate light is narrower than the probe wave's pulse width because P_{NL} is proportional to the product of the three pulses. Furthermore, the pulse wave shape of the phase-conjugate wave is symmetrical. This is the case for the usual process, and all of the discussions in the sections up to now have been in this category.

On the other hand, if the incident light pulse width is sufficiently narrower than the substance's response time, the polarization produced by the nonlinear interaction between the light wave and the substance cannot respond to instantaneous changes in the electric field where the magnitude of nonlinear polarization is proportional to the integrated energy of the incident light pulse — that is, proportional to the incident pulse width. If this is applied to degenerate four-wave mixing, we come to the conclusion that the pulse width of the phase-conjugate wave is approximately equal to that of the probe wave when the response time of the medium is sufficiently longer than the incident probe wave's pulse width. The nonlinearity obtained from measured values is then an effective value for the medium, so to speak, and is called the transient response. The resulting diffraction grating can be called a transient grating. We can study the relaxation process in a medium in detail by performing transient response analysis. Nonlinear polarization is actually an applied example of this phenomenon.

b. Transient Response in a Counterpropagating Arrangement

Transient response has been studied by assuming two-level atoms and an incident light pulse width sufficiently narrower than the response time of the medium.[20] Transient responses in a nonlinear medium for various pump waves and pulse times are shown in Figure 4.10.

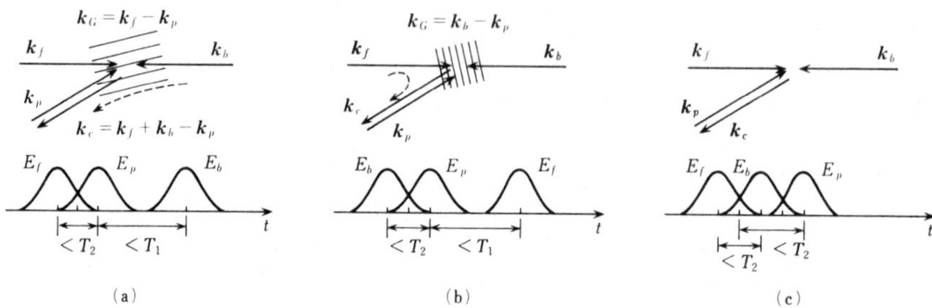

Figure 4.10 Transient response in four-wave mixing for pulsed incident waves. *(After Fujimoto and Yee. ©1986 IEEE)*[20]

(a) transmission diffraction grating, (b) reflection grating, (c) temporal diffraction grating
T_1: Energy (vertical) relaxation time T_2: Phase (horizontal) relaxation time k_G: Diffraction grating vector

In Figure 4.10(a), if the forward pump wave and the probe wave are superposed within the medium's phase (horizontal) relaxation time T_2, a spatial diffraction grating $k_G = k_f - k_p$ is formed due to the population difference between the upper and lower levels. Only if the back pump wave arrives at the same time as the two previously-mentioned light waves, or arrives afterward (but still within the range of the energy relaxation time T_1), is the back pump wave scattered to produce phase-conjugate light in the direction opposite that of the probe wave. On the other hand, in Figure 4.10(b), if only the back pump wave and probe wave are superimposed within the medium's horizontal relaxation time T_2, and the forward pump wave arrives simultaneously with the previously-mentioned two light waves — or afterwards — the forward pump light is scattered by the spatial diffraction grating $k_G = k_b - k_p$ to produce phase-conjugate light. The case of a wide grating spacing as in (a) corresponds to a slow relaxation process; a narrow spacing as in (b) to a fast relaxation process.

The nonlinear polarization corresponding to (a) is obtained by using the assumption $|A_c|, |A_p| \ll |A_f|, |A_b|$

$$P_{NL} \propto \int_{-\infty}^{t} dt_3 \, A_b(t_3) \exp[-(t-t_3)(1+i\delta)/T_2]$$

$$\int_{-\infty}^{t_3} dt_2 \exp[-(t_3-t_2)/T_1]$$

$$\int_{-\infty}^{t_2} dt_1 \{ A_f(t_2) A_p{}^*(t_1) \exp[-(t_2-t_1)(1-i\delta)/T_2] + \text{c.c.} \} \tag{4.57}$$

for light waves of frequency ω injected into two-level atoms (the atomic resonance frequency is ω_R).[21] The normalized frequency shift is $\delta = (\omega - \omega_R)T_2$. If the waves that form the spatial diffraction grating are injected simultaneously, since the diffraction grating relaxes with an energy relaxation time T_1, if the reflectance is measured while the pump light that should be scattered to become phase-conjugate light is delayed, then we can find T_1 from the rate of change of the reflectance.

As shown in Figure 4.10(c), there is a case where phase-conjugate light is generated even though the probe wave is injected after the two pumps. This is related to resonant two-photon absorption, and as shown in Figure 4.6, three-level atoms are used. This is explained as follows.[13] The forward pump and back pump waves counterpropagate within the phase relaxation time T_2. When the waves are injected into a nonlinear medium, they are spatially uniform; oscillating polarization emerges at twice the frequency of the injected waves. In addition, phase-conjugate light is produced if the probe wave is injected within the phase-relaxation time T_2. Consequently, if the pump waves are injected simultaneously, while the probe wave is delayed relative to the pump waves, then by determining the phase-conjugate light intensity I_c it is possible to estimate the medium's phase relaxation time T_2 from

$$I_c \propto \exp(-2\tau/T_2) \tag{4.58}$$

where τ is the delay time.

In general, since the phase (horizontal) relaxation time is shorter than the energy (vertical) relaxation time, the phenomenon in Figure 4.10(c) is less likely to occur than (a) and (b). Although these three processes are independent, there are also cases where mixtures are involved, depending on the medium.

c. Transient Response with Complex Relaxation Processes

When complex relaxation processes participate in the transient response of a nonlinear medium, the polarization corresponding to transmission and diffraction gratings can be written as

$$P_{NL} \propto A_b(t) \int d\tau\, W(t-\tau\,;\ \tau_{gT}) \int dt'\, A_f(t')\, A_p{}^*(t'+\tau_p)$$

$$+ A_f(t) \int d\tau\, W(t-\tau\,;\ \tau_{gR}) \int dt'\, A_b(t')\, A_p{}^*(t'+\tau_p) \tag{4.59}$$

if the effects of T_2 are neglected. Here, $W(t\,;\ \tau_g)$ is the medium response function. If each relaxation process has an attenuation constant as previously assumed then

$$W(t\,;\,\tau_g) = \exp(-t/\tau_g) \tag{4.60}$$

When the medium can be expressed as a two-level system, the relaxation time τ_g agrees with T_1.

d. Forward Transient Response

Forward transient responses for pump waves (electric field A_f, wave vector $\boldsymbol{k}_f$) and probe waves (electric field A_p, wave vector $\boldsymbol{k}_p$) propagating in the same direction have been studied based on nonlinear media modeled on two-level atoms.[22] When the frequency ω of both waves is near the medium's resonant frequency ω_R and they are injected into a medium within the phase relaxation time T_2, they produce an excitation wave that is proportional to the electric field product $A_f A_p{}^*$. Because of this, A_f self-diffracts in the $2\boldsymbol{k}_f - \boldsymbol{k}_p$ direction and phase-conjugate light is generated. In particular, when the medium has a uniform width, if the probe wave is delayed relative to the pump wave, we can find the medium's phase relaxation time T_2 from Equation (4.58) by determining the phase-conjugate light intensity I_c. When the medium has nonuniform width, $-2\tau/T_2$ in the expression for I_c may be replaced by $-4\tau/T_2$ in the exponent of Equation (4.58).

(2) Propagation Functions in Optical Kerr Media
a. Derivation of Fundamental Equations

As in Figure 4.11, a *continuous* pump wave (plane wave) E_f of frequency ω is injected into a nonlinear medium from the left side.

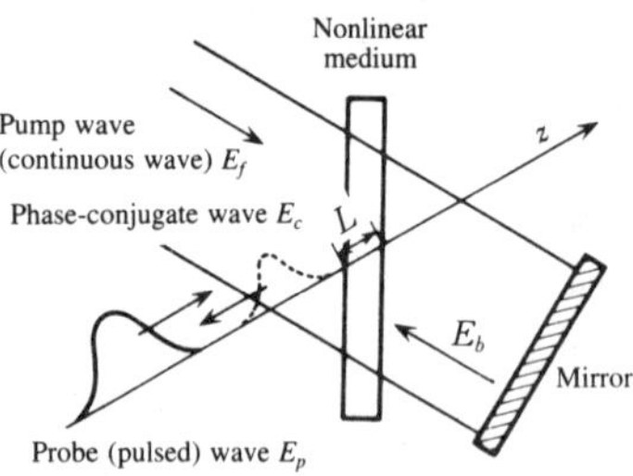

Figure 4.11 Transient response characteristics for beam positions. *(After Marburger)*[24]

Suppose that this wave is folded back onto the original wave path by a 100% reflective mirror. The reflected wave becomes E_b. Suppose further that a fourth signal light (actually, this is phase-conjugate light, as we will see later) E_c emerges if a third light wave E_p (a *pulsed* probe wave) of frequency ω is injected into the same medium. The probe wave propagates in the direction of the z axis and we assume that a nonlinear medium (having third-order nonlinearity

$\chi^{(3)}$ [m^{2}V^2]) occupies the region $0 < z < L$. The electric fields of these four waves are expressed as follows:

$$E_p(z, t) = (1/2)\, A_p(z, t)\exp[i\omega(t - z/v)] + \text{c.c.} \tag{4.61a}$$

$$E_c(z, t) = (1/2)\, A_c(z, t)\exp[i\omega(t + z/v)] + \text{c.c.} \tag{4.61b}$$

$$E_f(z, t) = (1/2)\, A_f\exp[i\omega(t - z/v)] + \text{c.c.} \tag{4.61c}$$

$$E_b(z, t) = (1/2)\, A_b\exp[i\omega(t + z/v)] + \text{c.c.} \tag{4.61d}$$

where the A_j (z,t) [V/m] are the spatially and temporally slowly-varying complex envelopes relative to the phase term, the A_j are the complex envelopes of the continuous waves, z [m] is the position, and v [m/s] is the phase velocity in the medium.

Assuming the nonlinear medium is nonmagnetic, if we consider the optical absorption in the medium, the wave equation for the nonlinear process can be written in the form (3.12). Among the third-order nonlinear polarizations formed by the above-mentioned pump, probe and phase-conjugate waves, those validly effecting the production of phase-conjugate light are similar to the two components of Equations (3.18a and b). Now suppose that the probe and signal waves are propagating in opposite directions. Substitute these in Equation (3.12). If we select the parts common to each component's frequency and wave-number terms, we obtain the coupled-mode equations for the probe and phase-conjugate electric fields.[23] That is,

$$\frac{\partial}{\partial z}A_p{}^* + \frac{1}{v}\cdot\frac{\partial}{\partial t}A_p{}^* + \frac{1}{2}\alpha_0 A_p{}^* = i\varkappa A_c \tag{4.62a}$$

$$\frac{\partial}{\partial z}A_c - \frac{1}{v}\cdot\frac{\partial}{\partial t}A_c - \frac{1}{2}\alpha_0 A_c = i\varkappa^* A_p{}^* \tag{4.62b}$$

$$\varkappa^* \equiv (3\omega\chi^{(3)}/nc)\, A_f A_b \tag{4.63}$$

where $\varkappa$ [m^{-1}] is the complex coupling coefficient, α_0[m^{-1}] is the absorption coefficient, and n is the medium's refractive index. If we assume that the electric field amplitudes of the two pump waves propagating in the medium are not attenuated, the coupling coefficient can be treated as a constant. When deriving the above equation, since the probe and signal amplitudes vary slowly in space and time, the adiabatic approximation

$$|\partial^2 A_j /\partial z^2| \ll |(\omega/v)(\partial A_j /\partial z)|$$

is applied to the spatial term and the adiabatic approximation

$$|\partial^2 A_j/\partial t^2| \ll |\omega(\partial A_j/\partial t)| \ll |\omega^2 A_j|$$

is applied to the temporal term. When there is no absorption ($\alpha_0 = 0$) and in the steady state ($\partial/\partial t = 0$), Equations (4.62a and b) revert to Equations (3.22a and b).

b. Transfer Functions

Consider the following complex Fourier transformation for solving Equations (4.62).

$$\tilde{f}(z, \Omega) = \int_{-\infty}^{\infty} F(z, t)\exp(-i\Omega t)\, dt$$

$$(4.64)$$

Fourier transforms are identified by the tilde ($\sim$). Ω is a complex number. If we use this transformation, Equations (4.62a and b) can be written as:

$$\frac{\partial}{\partial z}\tilde{a}_p{}^*(z, \Omega) + \left(i\frac{\Omega}{v} + \frac{1}{2}\alpha_0\right)\tilde{a}_p{}^*(z, \Omega) - ix\tilde{a}_c(z, \Omega) = 0 \qquad (4.65a)$$

$$\frac{\partial}{\partial z}\tilde{a}_c(z, \Omega) - \left(i\frac{\Omega}{v} + \frac{1}{2}\alpha_0\right)\tilde{a}_c(z, \Omega) - ix^*\tilde{a}_p{}^*(z, \Omega) = 0 \qquad (4.65b)$$

In order to solve Equations (4.65), we assume that the already-known probe wave of temporal waveform $F(t)$ is injected into the medium with boundary conditions $A_p(0,t) = F(t)$ and $A_c(L,t) = 0$. These conditions become $\tilde{a}_p{}^*(0,\Omega) = \tilde{f}{}^*(\Omega)$ and $\tilde{a}_c(L,\Omega) = 0$ in the Fourier domain. We obtain the following equation as the solution for the outgoing phase-conjugate light:[23]

$$\tilde{a}_c(z, \Omega) = H(z, \Omega)\tilde{f}{}^*(\Omega) \qquad (4.66)$$

$$H(z, \Omega) = \frac{-ix^*\sin[\beta(L-z)]}{\beta\cos(\beta L) + [i(\Omega/v) + (\alpha_0/2)]\sin(\beta L)} \qquad (4.67)$$

$$\beta \equiv \{x^2 - [(i\Omega/v) + (\alpha_0/2)]^2\}^{1/2} \qquad (4.68)$$

where $H(z,\Omega)$ is the transfer function.

The time-domain expression for phase-conjugate light is found using an inverse Fourier transform:

$$A_c(z, t) = \frac{1}{2\pi}\int_{-\infty+iy}^{\infty+iy} H(z, \Omega)\tilde{f}{}^*(\Omega)\exp(i\Omega t)\, d\Omega \qquad (4.69)$$

Here, we assume that the singularity in the complex plane is below y. If, in particular, we assume steady behavior in Equation (4.69), it can be treated as a normal Fourier transform since there is no extremum and $y = 0$. Consequently, Equation (4.69) means that the Fourier transform of an injected probe wave of arbitrary waveform and the outgoing phase-conjugate light are linked by a

Fourier transform through the transfer function shown in Equation (4.67). The Fourier transform is convenient for practical numerical solutions since a fast Fourier transform is used. Complex arguments can also be applied in the case of singularities.[23]

If we compare Equation (4.67) for steady-state detuning with Equation (3.47b), we see that both equations agree if $-|\delta k|/2 = \Omega/v$; this corresponds to the case where $z = 0$ and $\alpha_0 = 0$. In other words, there is a close relationship between the steady-state characteristics when there is detuning between the probe and pump waves and the transient response. This is explained as follows. In a normal Fourier transform, $\Omega = \omega - \omega_0$, and here $\tilde{f}^*(\Omega) = \tilde{A}_p^*(\Omega) = [\tilde{A}_p(-\Omega)]^*$. If we consider the Fourier component of Equation (4.69), we see that the frequency ω_0 of the pulsed probe wave is equal and opposite to that of the phase-conjugate wave.[24] This agrees with the steady-state characteristics with detuning. In addition, when there is detuning of the pump and probe waves the diffraction grating moves at a constant speed in response to the detuning (moving grating), and phase-conjugate light is produced. This movement corresponds to transient-response pulsed-probe movement (the pump is a continuous wave).

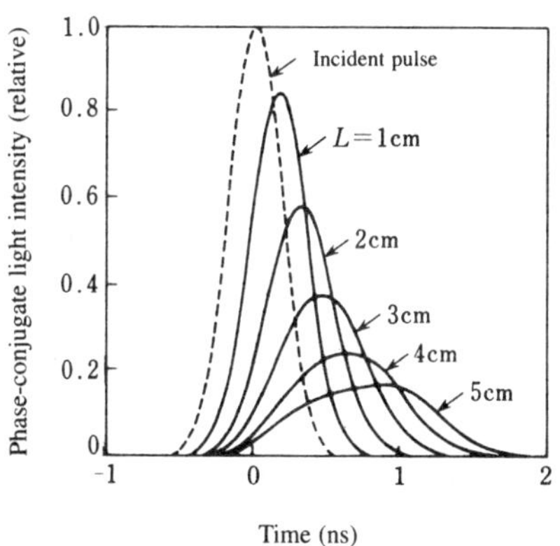

Figure 4.12 Waveforms for phase-conjugate light in an optical Kerr medium for various thicknesses. *(After Fisher, Suydam and Feldman)*[23]

The incident light pulse has width 0.5 ns (full width at $1/e$ times maximum), $\kappa L = \pi/4$, κ: Coupling coefficients, L : Thickness of test material. Test material is assumed to be Ge with a refractive index of 4.0.

Figure 4.12 shows the relationship between the thickness of the nonlinear medium and the response characteristics of phase-conjugate light.[23] As a numerical example, let the product of the coupling coefficient and the medium thickness $\kappa L = \pi/4$ be fixed, but allow L to vary. This means that the reproducibility of the temporal behavior of generated phase-conjugate light is better if the thickness of the nonlinear medium is thinner, given a fixed pulse width of the incident wave (a Gaussian waveform with $1/e$-folding time of 0.5 ns). If the medium is relatively thick, the temporal width of the phase-conjugate wave increases and the peak value falls, thus lowering the reflectance. Another way of looking at this is that it is necessary for the incident pulse width to be sufficiently larger than the transmission time through the nonlinear medium in order to obtain high reflectance. This result is explained as follows. Assume a nonlinear medium of constant thickness having a constant frequency-response bandwidth. If the pulse width of the incident light increases, the pulse frequency band narrows and the frequency components are efficiently converted to phase-conjugate light.

(3) Transfer Functions in a Photorefractive Medium

We will assume here that the carriers are electrons and will deal with the photorefractive effect through the band-transport model. As a result, the same constitutive relations and wave equation as in Section 3.3(3) can be used. Take the same four-wave mixing setup as in Figure 3.1. Also take counterpropagating forward E_f and back E_b pump waves and inject them into a nonlinear medium. Assume that the angle formed by the forward pump wave and the probe wave E_p within the crystal is θ, and further assume that the generated phase-conjugate light E_c emerges in the direction opposite that of the probe wave. Take the direction of the probe wave to be the z axis, and assume the nonlinear medium occupies the region $0 < z < L$. Assume also that the polarizations of these four waves are the same, and that their electric fields are represented as in Equations (4.61). The diffraction grating vector $k_G = k_f - k_p$ dominates, and we have a transmission grating. We assume that the electric fields of the two pump waves are of the first order, and that the electric field of the probe and phase-conjugate waves is of order Δ ($\Delta \ll 1$). Terms of order greater than Δ^2 are neglected.

In order to get linear constitutive relations, we consider the internal electric field created by space charges; the externally impressed electric field is taken as the sum of the newly-generated terms depending on the existence of the probe and phase-conjugate waves and the contribution when only the pump wave is present. We will use an adiabatic approximation. Assume an isotropic medium in which a change in the index of refraction is caused by the Pockels effect, and for which the wave Equation (3.63) is used. If pump depletion is neglected, only the terms satisfying $k_f + k_b = k_p + k_c$ are valid. Based on our supposition, we apply the complex Fourier transform of Equation (4.64) to the equation thus obtained, yielding[25]

$$\frac{\partial}{\partial z}\tilde{a}_p{}^*(z, \Omega) + i\left(\frac{\Omega}{v} - \Gamma\,|A_f|^2\right)\tilde{a}_p{}^*(z, \Omega) = i\Gamma\,(A_f A_b)^*\tilde{a}_c(z, \Omega)$$

$$(4.70a)$$

$$\frac{\partial}{\partial z}\tilde{a}_c(z,\Omega) - i\left(\frac{\Omega}{v} + \Gamma|A_b|^2\right)\tilde{a}_c(z,\Omega) = i\Gamma(A_fA_b)\tilde{a}_p^*(z,\Omega)$$

(4.70b)

where

$$\Gamma = \frac{-\omega n_0^3 r_{fp}}{(1/\tau_g) + i(\omega_g + \Omega)} \cdot \frac{E_{fp}^s}{(I_f + I_b)}\left(\frac{1}{\tau_g} + i\omega_g\right)$$

(4.71)

Here $v = c/n_0$ is the phase velocity in the medium; n_0 is the index of refraction for a non-oscillating system; r_{ij} is the opto-electric coefficient; and the I_j ($j = f, b$) are the optical intensities when the pump waves exist independently; τ_g is the diffraction grating rise time up to the point where the forward pump/probe interference term E_{fp} reaches the steady-state value E_{fp}^s; and ω_g is the oscillation frequency. These expressions are the same as those previously shown in Equations (3.82) and (3.83). Equation (4.70) is the fundamental equation for finding the transient response characteristics.

If we assume the waveform of the incident probe wave to be $F(z,t)$, the boundary conditions can be established as $\tilde{a}_p^*(0,\Omega) = \tilde{f}^*(0,\Omega)$ and $\tilde{a}_c(L,\Omega) = 0$. Equations (4.70a and b) can be solved based on these boundary conditions, and if the transfer function is defined as $H(z,\Omega) = \tilde{A}_c(z,\Omega)/\tilde{A}_p(z \to 0,\Omega)$, we obtain the following equation:[25]

$$H(0,\Omega) = \frac{-i\eta M}{\beta\cot(\beta L) + i[(\tau_c\omega_N/\tau_g L) + \rho M]}$$

(4.72)

where

$$\beta = \{[(\tau_c\omega_N/\tau_g L) + \rho M]^2 + (\eta M)^2\}^{1/2}$$

(4.73a)

$$\rho = [(r-1)/(r+1)](\xi/2)$$

(4.73b)

$$\eta = [2r^{1/2}/(r+1)](\xi/2)$$

(4.73c)

$$M = (1 + i\omega_g\tau_g)/[1 + i(\omega_N + \omega_g\tau_g)]$$

(4.73d)

$$\xi = -(\omega n_0^3/4c) r_{fp}|E_{fp}^s|\exp(-i\psi)$$

(4.73e)

$$r = I_b/I_f$$

(4.73f)

Here $\tau_c = L/v$ is the transit time through the medium, $\omega_N = \Omega\tau_g$ is the normalized frequency, ξ is the complex coupling coefficient, and r is the ratio of pump wave intensities. Various cases can be argued based on Equation (4.72).

Consider the steady state as an initial special case. Here, if $\omega_N = 0$ in Equation (4.72), and if $M = 1$ and $\beta = \xi/2$, we get

$$H(0,0) = \frac{-2r^{1/2}}{(r-1) - i(r+1)\cot(\xi L/2)}$$

(4.74)

This is equivalent to Equation (3.73b) for a photorefractive medium. If $r = 1$ (equivalent pump wave intensity) in the above equation, it agrees in form with Equation (3.25b) for a steady response in an isotropic optical Kerr medium.

When the diffraction grating rise time τ_g is sufficiently narrower than the transit time τ_c through the medium ($\tau_c / \tau_g \gg 1$), Equation (4.72) agrees with the transient response Equation (4.67) for an optical Kerr medium given a coefficient of absorption $\alpha_0 = 0$. Actually, we also obtain this from $r = 1$ and $\tau_g \to 0$ through Equation (4.72). Therefore, an optical Kerr medium can be regarded as the extreme condition where a photorefractive medium is excited by the same pump wave intensity, and where the medium responds in an infinitesimally short time. We have $\tau_c / \tau_g \ll 1$ for photorefractive media, with the exception of semiconductors. The transfer function here can be written as

$$H(0, \omega_N) = \frac{-2r^{1/2}}{(r-1) - i(r+1)\cot(M\xi L/2)} \tag{4.75}$$

The transient-response characteristics of phase-conjugate light can be found by using Equation (4.69) or

$$A_c(t) = \int_{-\infty}^{\infty} h(t - \tau) A_p^*(t)\, d\tau \tag{4.76}$$

Here $h(t)$ is the impulse response function, obtained by Fourier transformation of the transfer function $H(0, \omega_N)$. Figure 4.13 shows a calculated example of the impulse response function for an optical Kerr medium and a photorefractive medium.[25]

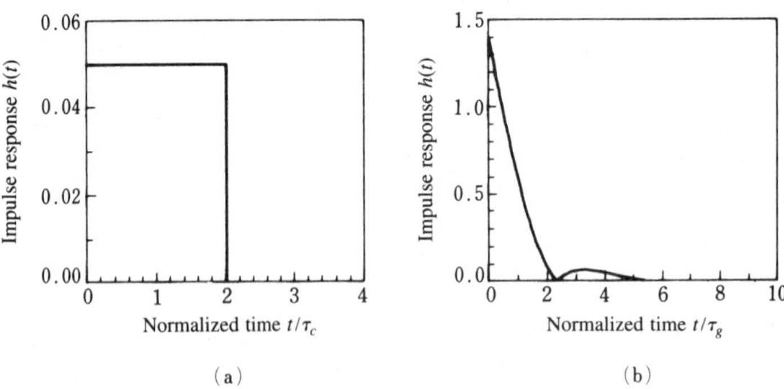

Figure 4.13 Impulse response of phase-conjugate light production process by four-wave mixing. *(After Papen, Saleh and Tataronis)* [25]

(a) Optical Kerr medium: $r = 1$, $\gamma L = 0.1$
(b) Photorefractive medium: $r = 0.5$, $|\gamma| L = 1$, $\psi = \pi/2$
τ_c: Transit time through medium, τ_g: diffraction grating rise time

Since the diffraction grating rise time τ_g is shorter for an optical Kerr medium than for a photorefractive medium, we get almost flat response characteristics for the transit time through the medium. On the other hand, since τ_g is relatively larger in a photorefractive medium, the response time is almost determined by τ_g. Here, when phase-conjugate light is generated in a slow-response-time photorefractive medium, in the strictest sense the outgoing wave shows a small shift due to the pump wave's complex conjugate.

5.

Principles of Phase-Conjugate Light Generation (Other Than Four-Wave Mixing)

This chapter introduces three-wave mixing and stimulated scattering as phase-conjugate light principles that do not depend on four-wave mixing.

5.1 Three-Wave Mixing

Three-wave mixing was first conceived of as a means for recovery from image degradation when optical images were sent through a multi-mode optical fiber. As shown in Figure 5.1, one pump wave E_L of frequency 2ω and one probe wave E_p of frequency ω are injected into a medium (length L) having second-order nonlinearity $\chi^{(2)}$. An idler wave E_I of frequency ω is obtained as a result of the parametric interactions. The idler wave becomes phase-conjugate to the probe under certain conditions. We next explain this through a set of equations.

Take the z axis as the direction of propagation of the pump wave. The electric fields of the probe, the pump and the idler waves are expressed as follows:

$$E_{jm}(\mathbf{r}, t) = (1/2) A_{jm}(\omega_j, \mathbf{r}) \exp[i(\omega_j t - \mathbf{k}_j \cdot \mathbf{r})] + \text{c.c.}$$

$$(j = L, p, I, \quad m = x, y \text{ or } z) \tag{5.1}$$

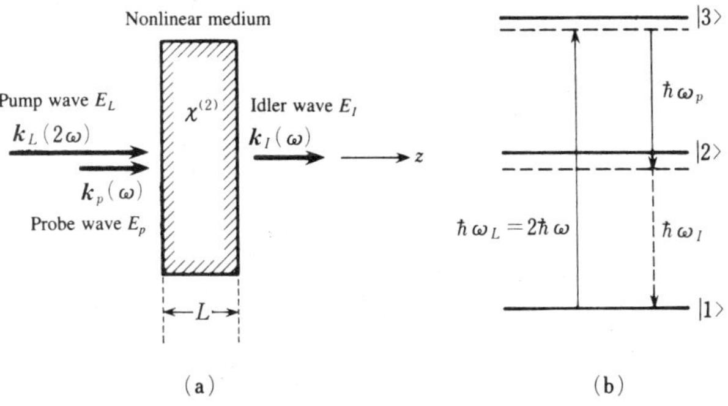

Figure 5.1 Optical configuration for three-wave mixing; three-level scheme.

(a) Optical configuration, (b) interaction of three-level atoms and light.
Phase-matching condition is $k_I(\omega) = k_L(2\omega) - k_p(\omega)$.

where the A_{jm} are the complex electric field amplitudes, ω_j is the frequency, k_j is the wave vector, $|k_j| = n_j\omega_j/c$ is the wave number, n_j is the index of refraction at frequency ω_j, c is the speed of light in vacuo and r is the position vector. If these waves exist in a $\chi^{(2)}$ medium, the nonlinear polarization is

$$P_{NLm} = \varepsilon_0 \chi^{(2)}_{mno} (\omega_I ; \ \omega_L, -\omega_p) A_{Ln}(\omega_L, r) A_{po}{}^*(\omega_p, r)$$
$$\cdot \exp\{i[(\omega_L - \omega_p)t - (k_L - k_p)\cdot r]\} + \text{c.c.} \quad (m, n, o = x, y \text{ or } z) \tag{5.2}$$

and yields the terms of proportionality $A_{pn}(\omega_p, r)A_{Io}(\omega_I, r)$ and $A_{Ln}(\omega_L, r)A_{Io}{}^*(\omega_I, r)$. In order to simplify what follows, we disregard the subscripts on the orthogonal components. We substitute these in wave Equation (3.12) with $\sigma = 0$, assuming that the electric field amplitude varies slowly, use the adiabatic approximation $|d^2A_j/dz^2| \ll |k_j(dA_j/dz)|$ and ignore the second derivative. Then we extract the equivalent frequency and wave-number terms to obtain the following parametric coupled equations:

$$\frac{dA_L}{dz} = -i\frac{2\omega\chi^{(2)}(2\omega)}{n_L c}A_p A_I \exp(i|\delta k|z) \tag{5.3a}$$

$$\frac{dA_p}{dz} = -i\frac{\omega\chi^{(2)}(\omega)}{n_p c}A_L A_I{}^* \exp(-i|\delta k|z) \tag{5.3b}$$

$$\frac{dA_I}{dz} = -i\frac{\omega\chi^{(2)}(\omega)}{n_I c}A_L A_p{}^* \exp(-i|\delta k|z) \tag{5.3c}$$

where $\delta k = k_p + k_I - k_L$ is the phase mismatch. Although we are using the phase-matching condition

$$k_I(2\omega - \omega) = k_L(2\omega) - k_p(\omega) \tag{5.4}$$

this is satisfied only for a constant direction in a specific medium. Here the $\phi_j(x,y)$ are the spatial data and $A_j(\omega_j, r) = |A_j|\exp[\phi_j(x,y)]$ are the complex amplitudes. If the parametric gain is small, and if the electric field of the pump is assumed not to vary in the medium, we obtain

$$A_I = i\eta(x, y) L |A_P| \exp\{i[\phi_L(x, y) - \phi_P(x, y)]\} \tag{5.5a}$$

$$\eta(x, y) = (2\pi/n_I\lambda)\chi^{(2)}|A_L(x, y)| \tag{5.5b}$$

as the idler wave amplitude after propagating distance L, where n_I and λ are the refractive index and wavelength of the idler.[1]

The following conditions are necessary in Equations (5.5) in order to obtain phase-conjugate light. (i) If there is no pump phase matching, the idler wave strictly becomes phase-conjugate to the probe. (ii) The pump and probe waves are polarized in the same direction as a phase-matching condition, and if the idler wave has orthogonal polarization, then it can be separately detected even though it propagates in the same direction.

Three-wave mixing has been proposed as a rigorous method for producing phase-conjugate light that is unaccompanied by a frequency shift. However, we must satisfy the problematic phase-matching conditions, making it necessary to provide the desired second harmonic as a pump wave. By using the $\chi^{(2)}$ process, the medium is restricted to those lacking inversion symmetry; there are other flaws as well. These flaws become a bottleneck to experimentation.

5.2 Stimulated Optical Scattering

Since phase-conjugate light was historically explained in terms of stimulated Brillouin backscattering,[2] an understanding of this phenomenon is of deep significance. After deriving the basic equations for stimulated Brillouin scattering, we will show that the cross-sectional internal electric field distribution of Stokes light emerging from stimulated Brillouin scattering is proportional to the complex conjugate of the cross-sectional distribution of the pump wave. Finally, we will explain the main points of stimulated Raman scattering.

(1) Stimulated Brillouin Scattering
a. Fundamental Equations

Stimulated Brillouin scattering (SBS) is a phenomenon wherein only frequency-shifted components of injected light and sound waves are generated through the interaction of light waves with sound waves and acoustical lattice oscillations (phonons), thus producing strong backscattering. Here the frequency shift ω_B is called a Brillouin shift, the low-frequency scattered light is called Stokes light, and the high-frequency scattered light is called anti-Stokes light. We next explain Stokes light mathematically.

The light waves are shown in Figure 5.2, with the z axis taken as the direction of propagation of the pump wave.

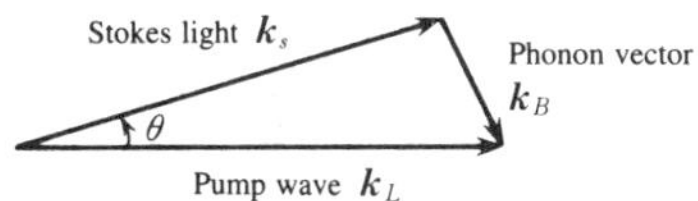

Figure 5.2 Principle of stimulated Brillouin scattering.

Pump wave that is injected into the medium exits at scattering angle θ as Stokes light due to the movement of sound waves (phonons) at velocity ν_a.
$k_s = k_L - k_B$, $|k_B| = \omega_B / \nu_a$

Now we represent the electric fields of the pump wave E_L and the Stokes wave E_s as

$$E_j(r, t) = (1/2) A_j(r_\perp, z) \exp[i(\omega_j t - k_j \cdot r)] + \text{c.c.} \quad (j = L, s) \tag{5.6}$$

where the $A_j(r_\perp, z)$[V/m] are the complex electric field amplitudes, ω_j is the wave frequency, k_j is the wave vector with $|k_j| = n_j \omega_j / c$, n_j is the refractive index of the medium at frequency ω_j, c is the speed of light in vacuo, r is the position vector, and $r_\perp$ is the vertical cross section taken in the forward direction. Phonons are produced by fluctuations in the density of the medium, where the medium density ρ [kg/m³] is given by

$$\rho(r, t) = (1/2) \rho_B \exp[i(\omega_B t - k_B \cdot r)] + \text{c.c.} \tag{5.7}$$

Now the density of the medium satisfies the following equation.[3]

$$\frac{\partial \rho}{\partial t} + \Gamma_B \rho = i C_B E_L E_s{}^* \tag{5.8}$$

where Γ_B is the damping constant, $C_B = \pi n_L{}^5 p \varepsilon_0 / 2\lambda \nu_a$, p is the elasticity coefficient, ε_0 is the permittivity of free space, λ is the pump wavelength, and ν_a is the velocity of the acoustic wave. In addition, nonlinear polarization P_{NL} is produced by the interaction of the electric field E and the phonons, and can be written as

$$P_{NL} = 4\varepsilon_0 (\partial \varepsilon / \partial \rho)_T \rho E \tag{5.9}$$

Here ε is the permittivity and T is the temperature. We solve Equation (5.8) and, after substituting it in wave Equation (3.12), select the terms that satisfy the law of conservation of energy and the phase-matching conditions:

$$\omega_L - \omega_s = \omega_B, \qquad \mathbf{k}_L - \mathbf{k}_s = \mathbf{k}_B \tag{5.10a,b}$$

In experiments on the production of phase-conjugate light, since the difference in scattering angle θ from $180°$ is at most 3×10^{-3} radians,[2] we will take it to be $180°$. We thus obtain the equations for Stokes light:

$$\frac{\partial A_s}{\partial z} + \frac{i}{2|\mathbf{k}_s|} \Delta_t A_s + \frac{1}{2} g_B (n_L c \varepsilon_0/2)|A_L|^2 A_s = 0 \tag{5.11}$$

$$g_B = 2\pi n_L{}^7 p^2 / c\lambda^2 \rho_0 v_a \delta\nu_B \tag{5.12}$$

Here Δ_t is the horizontal Laplacian, g_B [m / W] is the Brillouin gain coefficient, ρ_0 is the average density, and $\delta\nu_B = \Gamma_B/\pi$ is the spectral width due to damping. Since the pump wave has a greater intensity than the Stokes wave, the following equation comes about if we ignore pump wave depletion and gain.

$$\frac{\partial A_L}{\partial z} - \frac{i}{2|\mathbf{k}_L|} \Delta_t A_L = 0 \tag{5.13}$$

We also make use of the slowly varying envelope approximation in solving the above equation.

$$|\partial^2 A_j/\partial z^2| \ll |\mathbf{k}_j(\partial A_j/\partial z)|$$

Equations (5.11) and (5.13) are the fundamental equations of stimulated Brillouin scattering.

b. Proof of Phase-Conjugate Light

From Equation (5.10b), the Brillouin shift ω_B is expressed by

$$\omega_B = \omega_L - \omega_s = 2\omega_L[n_L v_a/c]\sin(\theta/2) \tag{5.14}$$

In reality, since ω_B is of the order of 0.01–1 cm^{-1}, $(k_L - k_s)/k_L \approx 10^{-5}$ in the optical domain. Consequently, since the magnitudes of k_L and k_s can be regarded as being approximately equal, we assume below that $k_L \approx k_s = k$.

Now assume the expression for a Stokes wave is a first-order coupling function, and consider the following equations for a wave propagating in the $-z$ direction. Assume further a functional system $f_q(r_\perp, z)$ that satisfies the orthogonality conditions[2]

$$\frac{\partial f_q}{\partial z} + \frac{i}{2k} \Delta_t f_q = 0 \tag{5.15a}$$

$$\int f_p{}^*(r_\perp, z) f_q(r_\perp, z)\, dr_\perp = \delta_{pq} \tag{5.15b}$$

We begin by selecting the pump wave $A_L(r_\perp, z)$ as a multiple of $f_0{}^*(r_\perp, z)$. That is, we assume that

$$A_L(r_\perp, z) = C f_0{}^*(r_\perp, z) \tag{5.16}$$

C is a constant. Here, the remainder constant $f_q(r_\perp, z)$ ($q = 1, 2, \ldots$) is arbitrarily selected from orthogonality conditions (5.15b). The Stokes light is the following f_q series expansion:

$$A_s(r_\perp, z) = \sum_{q=0}^{\infty} C_q(z) f_q(r_\perp, z) \tag{5.17}$$

Next we can find the equation that the coefficient $C_q(z)$ must satisfy. We substitute Equations (5.16) and (5.17) in Equation (5.11) to obtain:

$$\sum_q \frac{dC_q}{dz} f_q + \frac{1}{2} g_B C^2 |f_0|^2 C_q f_q = 0 \tag{5.18}$$

After multiplying the above equation by $f_p^*(z)$, we integrate over a perpendicular cross section at z. If we use orthogonality conditions (5.15b), we obtain the following equation:

$$\frac{dC_p}{dz} + \frac{1}{2} \sum_{q=0}^{\infty} g_{pq}(z) C_q(z) = 0 \tag{5.19}$$

where

$$g_{pq}(z) = g_B C^2 \int dr_\perp |f_0(r_\perp, z)|^2 f_p^*(r_\perp, z) f_q(r_\perp, z) \tag{5.20}$$

In Equation (5.20), if the pump intensity distribution $|f_0(r_\perp, z)|^2$ fluctuates within a cross section, the $f_p^* f_q$ terms ($p \neq q$) cancel out to a certain extent, and a strong influence extends only to g_{00} rather than any other g_{pq}. As a result, the coefficient C_0 becomes dramatically large along with the propagation, allowing the Stokes light in Equation (5.19) to be described in essence by only the first term. Here, by using Equation (5.16), we obtain

$$A_s(r_\perp, z) \approx C_0(z) f_0(r_\perp, z) = A_L^*(r_\perp, z) C_0(z) / C^* \tag{5.21}$$

as the electric field of the Stokes light.[2] This means that the internal cross-sectional electric field distribution of the Stokes light $A_s(r_\perp, z)$ appearing through stimulated Brillouin backscattering is proportional to the complex conjugate of the pump wave's internal cross-sectional electric field distribution $A_L(r_\perp, z)$, thus satisfying the appropriate phase-conjugate light conditions.

We made several assumptions when we derived this effect. One of them was to disregard the Stokes-light/pump-wave frequency shift. In reality, one cannot obtain phase-conjugate light in the strictest sense of the word through SBS because of the previously-described shift. In addition, we assumed that the pump wave intensity distribution fluctuates within the cross section. However, in addition to g_{00}, g_{qq} ($q \neq 0$) is present when a pump wave with small fluctuations is used, and again we cannot obtain phase-conjugate light in the strictest sense of the word. This is significant for phase compensation experiments since pump waves passing through astigmatic particles can be used. The quality of phase-conjugate light emerging via stimulated light scattering in optical wave guides has also been discussed.[4] According to this argument, in order to produce high-quality phase-conjugate light, the product of the waveguide length, the Stokes shift and the excitation mode number divided by the cross-sectional area must be either small or have a constant value.

(2) Stimulated Raman Scattering

Stimulated Raman scattering (SRS) is a phenomenon that generates light waves whose frequencies are shifted from the incident frequency due to the interaction of light waves, molecular oscillation and optical grating oscillation (optical phonons). SRS approximates SBS in form. If we assume forward scattering, there is a resemblance to Equations (5.11) and (5.13) for SRS Stokes waves and pump waves. However, the symbols representing the second and third terms may be reversed in the Stokes light equation and the gain coefficient may be replaced by the Raman gain coefficient g_R. However, the frequency shift ω_R in SRS (Raman shift) is 3 to 4 orders of magnitude larger than the Brillouin shift ω_B, and the quality of the phase-conjugate light is not good because $(k_L - k_s)/k_s \approx 0.1$. It has also been shown for SRS that approximately phase-conjugate light can be obtained under certain conditions by using methods similar to those for SBS.[5]

6.
Experiments in Phase-Conjugate Light Generation

Phase-conjugate light is produced mainly by degenerate four-wave mixing. As we saw in Section 3.1, the advantages due to this are that phase-matching conditions are easily satisfied, that light from the same optical source can be used, and that there are no restrictions on the type of medium. In addition, any of a number of mechanisms described below can be used for the production of phase-conjugate light. We will introduce experimental techniques for generating phase-conjugate light in this chapter. Section 6.1 explains the characteristics of various experimental methodologies. Techniques common to all types of experiments, as well as optical setups, are explained in Section 6.2. Four-wave mixing using standardized methods for the production of phase-conjugate light will be explained in Section 6.3, where experimental data characteristics of each individual medium will also be shown. In Section 6.4, phase-conjugate light generation due to two-photon absorption resonance, photon echoes and other special mechanisms will be explained. The characteristics of four-wave mixing — particularly when narrow light pulses are injected into a medium — will be explained in Section 6.5. Section 6.6 will touch upon three-wave mixing experiments, and Section 6.7 will address experiments with stimulated optical scattering.

6.1 Introduction

As explained in Chapters 3 – 5, phase-conjugate light can be produced by a variety of mechanisms. It will be necessary to classify the methods of generation by objective in order to understand their characteristics as physical phenomena. Table 6.1 shows the characteristics of phase-conjugate light generation resulting from various techniques. Degenerate four-wave mixing making use of the optical Kerr effect is applicable if we assume the objective to be fast response. In order to produce phase-conjugate light by using comparatively low optical power, degenerate four-wave mixing using the photorefractive and band-filling effects is used. In general, fast response is incompatible with high nonlinearity.

Table 6.1 Characteristics of all types of phase-conjugate light generating methods

Generating method (Physical phenomenon)	Characteristics
Degenerate four-wave mixing (Optical Kerr effect)	$\chi^{(3)}$ process, same frequency for all four waves, no frequency shift between probe and phase-conjugate waves, no restrictions on medium symmetry, amplification is possible (reflectance > 1), relatively high response speed
Degenerate four-wave mixing (Photorefractive effect)	Equivalent $\chi^{(3)}$ process (including Pockels effect), same frequency for all four waves, no frequency shift, amplification is possible (reflectance > 1), can have comparatively low optical power
Thermal effects	Values comparatively high for $\chi^{(3)}$ process, response time is comparatively slow
Forward degenerate four-wave mixing	Produced by self-diffraction, thin test material is required because of transmittivity
Two-photon absorption	$\chi^{(3)}$ process, same frequency for all four waves, no frequency shift, interatomic level coherence
Photon echoes	Short wave pulses are essential, non-linear coherent effect
Three-wave mixing	$\chi^{(2)}$ process, phase matching is difficult, medium lacking inversion symmetry is required
Stimulated Brillouin scattering	Experiments became the origin for the discovery of phase-conjugate light, $\chi^{(3)}$ process, frequency shift, reflectance does not exceed 1
Stimulated Raman scattering	$\chi^{(3)}$ process, has a frequency shift, reflectance cannot exceed 1
Brillouin-enhanced four-wave mixing	Amplification is possible (reflectance > 1), $\chi^{(3)}$ process, has frequency shift

6.2 Basic Techniques for Producing Phase-Conjugate Light

We will outline basic techniques used to produce phase-conjugate light in this section. After explaining the types of optical setups, we will describe experimental apparatus for detecting phase-conjugate light.

(1) Types of Optical Configurations

The setups for producing phase-conjugate light by four-wave mixing can be classified as

(i) back degenerate four-wave mixing, and

(ii) forward degenerate four-wave mixing

depending on the propagation direction of the phase-conjugate wave relative to the direction of the probe wave. The back type is a standard experimental technique that emerged from the early stages of four-wave mixing experiments, and can further be divided into counterpropagating-pump, retroreflection, and self-pumped geometries. Although the forward type appears to be three-wave mixing in which two beams are injected and one beam emerges, it is classified as four-wave mixing since the phase-conjugate light is produced by four photons. We explain the application of various setups below.

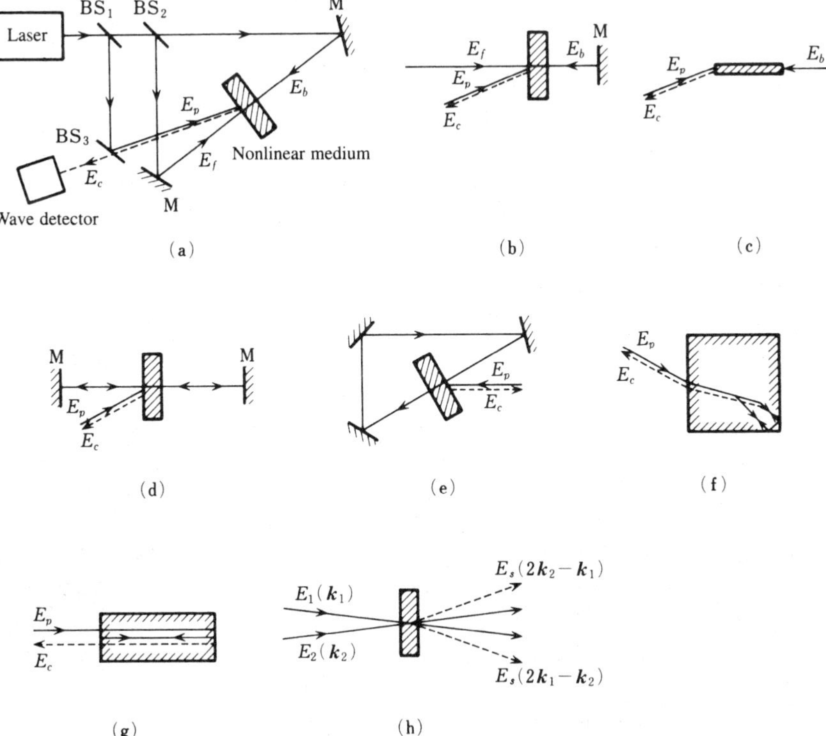

Figure 6.1 Types of optical setups for degenerate four-wave mixing.

(a) counterpropagating, (b) retroreflective, (c) retroreflection with optical fiber, (d) self-pumped (Fabry-Perot etalon), (e) self-pumped (ring resonator), (f) self-pumped (internal medium reflectance), (g) internal cavity, (h) forward wave

a. Counterpropagating-Pump Configuration

The basic setup for generating phase-conjugate light by degenerate four-wave mixing is shown in Figure 6.1(a); this is the counterpropagating-pump configuration. The wave coming from the optical source is divided into pump and probe waves by beam splitter BS_1. The pump wave is further separated into two wave paths by a separate beam splitter BS_2, becoming pump waves E_f and E_b. These pump waves counterpropagate from both sides of a nonlinear medium and are injected into it. On the other hand, probe wave E_p is reflected by beam splitter BS_3, and is injected into the nonlinear medium along with the pump waves. The resulting phase-conjugate wave E_c is guided to the observation system by beam splitter BS_3. The difference in distance from the light source to the nonlinear medium is held constant to within the coherence length of the laser beam. In particular, when pulsed light is used it is very important for the three beams all to be present at the same time.

b. Retroreflection Configuration

As shown in Figure 6.1(b), the pump wave is reflected back along its original path by a high reflectance, thus simplifying the optical system if a single pump-wave beam is assumed. Since the parts near the light source in (b) are common to those in (a), we have omitted these parts from the figure. When optical fiber is used for the nonlinear medium in (c),[1] this too can be looked upon as a type of retroreflection since the reflected waves in the optical fiber are used as pump waves.

c. Self-Pumped Configuration

When a nonlinear medium is used to achieve high internal gain — even when independent pump waves are not used — a self-induced grating is formed by the scattered light, the phase-matching conditions are automatically satisfied, and phase-conjugate light is produced. This is called the self-pumped configuration. As shown in Figure 6.1(d and e), the optical arrangement comprises a Fabry-Perot etalon and a ring resonator with an external mirror; (f) shows internal medium reflection. It is, however, necessary to adjust the optical system to allow for the previously separate rays when an external mirror is used to form the resonator. These configurations use $BaTiO_3$, CS_2, and SBN, among other substances. The internal-reflection medium is of course limited to solids.

d. Internal Cavity Configuration

As shown in Figure 6.1(g), a Fabry-Perot etalon with both ends of the non-linear medium acting as reflective surfaces is used for the special objectives of semiconductors and bistable optical operation. Here, the pump wave is multiply reflected within the resonator.

e. Forward-Type Degenerate Four-Wave Mixing

The configuration shown in Figure 6.1(h) shows strong absorption; it is not suitable for producing phase-conjugate light with a thin material. Intense pump waves E_1 and E_2 (wave-number vectors k_1 and k_2, respectively) excite the two forward waves. New light waves are then produced by interaction of these excited waves with one of the pump waves in the direction $k_s = 2k_1 - k_2$ (and $2k_2 - k_1$) that satisfies the phase-matching conditions. A pulsed light source is used to form transient-grating polarization; this technique is highly applicable to spectroscopy.

(2) Experimental Apparatus for Phase-Conjugate Light Detection

A high reflectance is of course very important for the production of phase-conjugate light. This not only is true for detecting and observing phase-conjugate light, but is also an important factor in applications. The following are considered for increasing reflectance, as indicated by Equations (3.28) and (3.23):

(i)   increase the nonlinear coefficient of susceptibility $\chi^{(3)}$;

(ii)   increase the pump wave intensity;

(iii)   increase the interaction length;

In particular, of great importance in the case of photorefractive media is

(iv)   the phase difference between the optical-intensity distribution and the refractive-index distribution.

Here, as explained in Section 3.3(1), the lattice spacing Λ of the diffraction grating also exerts great influence on the reflectance.

Because materials research is so essential for increasing nonlinearity, it will be mentioned in Chapter 7. There are two methods for increasing pump-wave intensity. One of them uses Q-switch and mode synchronization, where the optical power is increased by pulsing the optical source. Increasing the power of the optical source is not necessarily characteristic of phase-conjugate light techniques, and since the problems are similar to those with laser optics, we exclude that method here. A second method is waveguiding and, since there is an increase in power concentration, the effects are indistinguishable from increases in interaction length, as will be mentioned later.

The importance of this to an optical configuration is that the difference in distance between each beam source and the phase-conjugate wave-generating medium must be held constant to within the coherence length of the beam source. In order to do this, the coherence length of the source is increased; that is, both vertical and horizontal single modes are assumed. In the case of degenerate four-wave mixing, the phase-conjugate light is in the direction opposite that of the probe wave. In other words, light is prevented from being injected into the source since propagation is in a direction toward the source. This also makes

(v)   stable operation of the optical source, and

(vi)   experimental techniques for detection

very important.

Methods for detection of phase-conjugate light with experimental apparatus are shown in Table 6.2; these various techniques are explained below.

Table 6.2 Experimental processes for detecting phase-conjugate light

Item	Means	Specific measures
Increase in reflectance	Increase in nonlinearity	Make use of resonance effect (investigation of materials)
	Increase pump wave intensity	Pulse action of light source Nonlinear-medium waveguiding
	Increase interaction length	Nonlinear-medium waveguiding Nearly-linear arrangement
	Regulation of phase relationship	Moving diffraction grating
Stable operation of light source (single vertical and horizontal modes)	Reflected waves do not return to light source	Optical isolator immediately after light source
		Wave delay circuit immediately after light source
Simplification of detection	Improvement of signal-to-noise ratio	Polarization discrimination
		Frequency discrimination
		Time-domain discrimination
		Self-correlation method

a. Resonance Phenomena in Nonlinear Media

Resonance effects between energy levels can be used to increase nonlinearity in atomic, molecular, and semiconductor media. The optical Kerr effect is produced by a shift in energy levels which depends on optical intensity. This energy level shift brings about a shift in the atomic-resonance frequency (a change in the absorption bands), which has the effect of changing the refractive-index distribution. The resonance effect can be used with metallic vapors (Na, Rb, etc.), CuCl bi-excitons, semiconductor band gaps, and organic dye compounds.

As a specific example, we will explain the case where Na metal vapor is used. A 3×10^{20} m^{-3} concentration of Na vapor was sealed in a cell together with a 5-Torr He buffer atmosphere, into which a laser beam was injected (an N_2-excited dye laser). Figure 6.2 shows the measured phase-conjugate wave intensity versus the exciting wavelength.[2] Wavelengths of 589.6 nm and 589.0 nm correspond to the D_1 and D_2 resonance bands of Na, which account for the zeroes in the intensity curves. The location of resonance on the high-

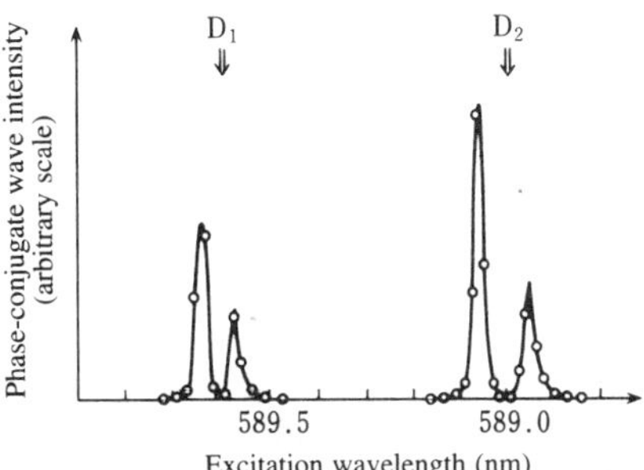

Figure 6.2 Phase-conjugate wave intensity for Na atmosphere. *(After Bloom, Liao and Economou)*[2]

Pump wave intensity: approximately 50 kW/cm^2
Probe wave intensity: approximately 500 W/cm^2
Na density: 3×10^{20} m^{-3}

frequency side is asymmetric in order to limit nonlinear optical interactions as a function of self-focusing. $n_2 \approx 1.1 \times 10^{-8}$ esu (1.2×10^{-17} m^2/V^2) is reached only at a frequency which is 1.25 cm^{-1} lower than the Na D$_1$ resonance band.

b. Waveguiding in Nonlinear Media

There exist methods for increasing reflectance in which the interaction length is increased through nonlinear waveguiding. Waveguiding also has the advantage of nonlinear effects occurring at low incident wave power because it brings about higher optical power concentrations than those in bulk materials. However, as the interaction length increases, waveguiding can be optimized only in a medium with a low absorption coefficient since the pump and probe wave intensities as well as the phase-conjugate wave intensity are attenuated by absorption. These circumstances can be thought of in a somewhat more quantitative manner.

When Gauss-distributed laser light (spot size: ω_0) converges through a lens on bulk material, the beam broadens after it converges. If we define the interaction length l as the distance to where spot size is $\sqrt{2}w_0$, then we obtain $l = 2\pi\omega_0^2/\lambda$, where λ is the wavelength of the light. On the other hand, if the length of the wave guide is L and the absorption coefficient is α, then we find

$$l_{\text{eff}} = [1 - \exp(-\alpha L)]/\alpha \tag{6.1}$$

as the effective interaction length of the waveguide.[3] The length of the waveguide can be approximated by $l_{\text{eff}} \approx L$ at small L, and by $l_{\text{eff}} \approx 1/\alpha$ at larger L. Consequently, when L is large

$$l_{\text{eff}}/l = \lambda/(2\pi w_0^2 \alpha) \tag{6.2}$$

becomes the interaction-length ratio. For example, if we assume a loss of 1 dB/km, then $1/\alpha = 4.34$ km, yielding an extreme interaction length of the order of kilometers. Moreover, if $\lambda = 1$ μm and $w_0 = 4$ μm, then $l_{\mathrm{eff}}/l = 4.3 \times 10^7$, showing us that we get an interaction length approximately 10^7 times greater due to waveguiding. Therefore, high nonlinear polarization is obtained from the effects of distance even with low nonlinearity. However, this technique is restricted by the materials from which the waveguide can be made. In addition, the optical-source coherence length must be longer than l_{eff}, and it is essential that both the vertical and horizontal waves have individual sources. The waveguide can be of (i) fiber or (ii) planar type.

An example of a liquid fiber is a waveguide containing CS_2 sealed in a glass tube. CS_2 is a non-resonant liquid whose absorptance $\alpha = 0.001$ cm^{-1} is unusually small. At the same time, good use can be made of its comparatively high refractive index $n_2 \approx 1.2 \times 10^{-11}$ esu (1.3×10^{-20} m^2/V^2). From Equation (6.1) the effective interaction length can be a maximum of about 10 m. This experimental example can be explained as follows.[4] Vertical and horizontal single-mode Ar ion lasers (wavelength: 514.5 nm) operate continuously, and cause counterpropagating degenerate four-wave mixing. The optical fiber has a core diameter of 4 μm, and the CS_2-filled fiber is 3 m long. Pressurized nitrogen gas (4100 Torr) at the injection end of the fiber prevents the liquid from leaking out. An objective lens is used to couple the beam to the optical fiber. The probe wave is coupled to the forward pump wave by the same lens. The input power of the pump wave to the fiber was 6 mW. The reflectance of the phase-conjugate light was 0.45%. Since the core diameter was an extremely small 4 μm in this experiment, the optical power concentration was increased; this produced phase-conjugate light of the order of milliwatts.

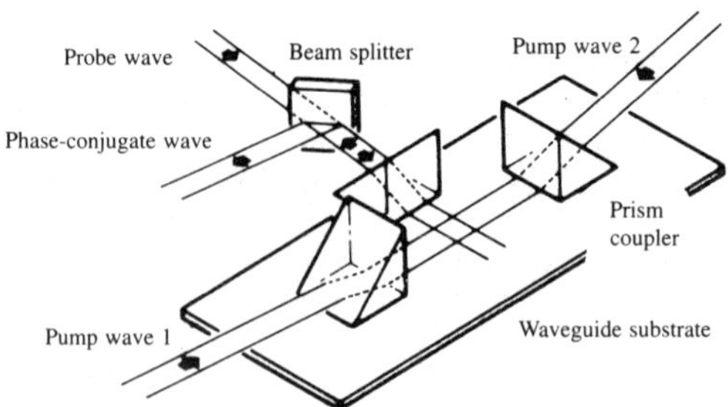

Figure 6.3 Experimental setup for degenerate four-wave mixing with planar waveguiding. *(After Gabel, Delong, Seaton and Stegeman)*[5]

The waveguide here is produced via ion exchange.

As an example of a planar waveguide, ions were exchanged on semiconductor-doped glass, making a simple-mode waveguide.[5] The substrate was a Corning 3-69 dyed glass filter (with CdS_xSe_{1-x} particles). When this was soaked in a KNO_3 solution for 30 hours at $400°C$, the K ions in the solution exchanged with the Na ions in the substrate, forming the waveguide. The refractive-index distribution in the waveguide varied exponentially as a function of the depth from the substrate surface. The change in the refractive index at the surface was 0.008 for a $1/e$ depth of approximately 3 μm. A frequency-doubled Q-switched Nd:YAG laser (wavelength: 0.532 μm; pulse width: 15 ns) was used as the light source; degenerate four-wave mixing was implemented by prismatic coupling, as shown in Figure 6.3. The probe wave was coupled orthogonally to the pump waves. Probe-wave entry was from the phase-conjugate side to minimize pump-wave scattering in the waveguide. The maximum reflectance obtained was 1%. This type of planar waveguiding was also attempted using organic dyes.

c. Nearly-Collinear Configuration

One method of attempting to increase the interaction length is to arrange the four light waves in degenerate four-wave mixing in a nearly collinear fashion. This is similar to the previously-mentioned arrangement used with multiple-mode optical fiber. This technique increases the reflectance for phase-conjugate light, and is effective when the intensity of the phase-conjugate light is somewhat higher than the intensity of the pump wave reflected by the surface of the medium. However, if the refractive index n of a nonlinear medium is high, the reflectance r will also be high. For example, when $n = 2.5$, r can be as high as 0.2. Therefore, when the reflectance is low, reflected pump light from the surface of the medium cannot be disregarded, making it difficult to distinguish between the two waves. In this case, we can consider "marking" the pump probe wave for distinction. As will be mentioned later, polarization, frequency, and time domain are used to make this distinction.

d. Moving Diffraction Grating

When given a photorefractive medium, we can expect a very important role to be played by the phase difference between the optical-intensity distribution and the refractive-index distribution vis-à-vis coupling between the pump and probe waves. Here, an electric field can be impressed on a crystal to increase the reflectance, and the interference bands can be moved spatially to maintain appropriate phase relationships. Since the diffraction grating is moving at this time, it is called a moving grating. The frequency-shift techniques for moving the diffraction grating are (i) sawtooth pump-wave oscillation by a piezo-element mirror followed by a Doppler shift, and (ii) frequency shift by an acoustic optical modulator. Section 6.3(5) deals with BSO crystal examples in detail.

e. Opto-Isolator Placed Immediately after the Optical Source

In the ideal optical configuration for degenerate four-wave mixing, the phase-conjugate wave propagates in a direction exactly opposite that of the probe wave. Consequently, the phase-conjugate wave that is produced could return to the light source, raising the risk of an unstable light source. In order to prevent this, an optical isolator is placed immediately after the source. The opto-isolator comprises two polarizers with axes making an angle of 45° and one Faraday rotator between them. The light beam coming from the source is linearly polarized after passing through the first polarizer; the axis of polarization is rotated 45° by the Faraday rotator, and the polarized light passes through the second polarizer. If the direction of propagation is reversed, the Faraday rotator causes the phase-conjugate light to have its polarization at 90° to the axis of the polarizer nearer the source, so that it does not return to the source. An opto-isolator is essential when the source oscillates in the vertical single mode.

f. Optical Delay Circuit Immediately after the Source

An optical delay circuit is used mostly with a pulsed wave source. The objective is stable operation of the light source, the same as the role mentioned above for the opto-isolator. As with phase-conjugate light returning to the light source when an outgoing pulse is not present, this technique adjusts the distance L between the source and the phase-conjugate medium according to the pulse repetition rate. In the case of air, a 1-ns delay in the round trip is used, corresponding to L = 15 cm.[6]

g. Polarization Discrimination

As previously mentioned, polarization discrimination can be used in cases where the reflectance is small in nearly-collinear configurations. When the polarization of a nonlinear medium is described by a tensor, phase-conjugate light may be detected even by varying the polarization of the pump and probe waves. The difference in polarization between the pump and probe waves is used to distinguish between the pump wave reflected from the medium surface and the phase-conjugate wave. This technique is called polarization discrimination. Since the probe and the phase-conjugate waves follow the same optical path, their polarizations are the same; orthogonal pump-wave polarization is ideal. However, in reality, when the polarizations of the phase-conjugate and pump waves are nearly orthogonal, optimization is made possible by completely isolating the reflected pump wave and sacrificing part of the phase-conjugate light.

We next show an actual example of the polarization discrimination technique. Assume an isotropic cubic crystal (point group: 432, $\overline{4}$3m, m3m, etc.). We can represent the third-order nonlinear polarization (complex amplitude) defined by Equation (C.1) and Table B.1 in the Appendix as

$$P = 3(\chi^{(3)}_{xxyy} + \chi^{(3)}_{xyxy})(A \cdot A^*)A + 3\chi^{(3)}_{xyyx}(A \cdot A)A^* \tag{6.3}$$
$$+ 3(\chi^{(3)}_{xxxx} - \chi^{(3)}_{xxyy} - \chi^{(3)}_{xyxy} - \chi^{(3)}_{xyyx})\,\text{Tr}(AA) \cdot A^*$$

The subscripts represent the crystal's principal axes.

If we assume the counterpropagating pump configuration of Figure 6.1(a) as the setup for degenerate four-wave mixing, we can write the contribution polarization makes to the production of phase-conjugate light as:

$$P_c^{(3)} = A[(A_f \cdot A_p^*)A_b + (A_b \cdot A_p^*)A_f] + 2C(A_f \cdot A_b)A_p^*$$
$$+ (B-C)[\text{Tr}(A_f A_b) \cdot A_p^* + \text{Tr}(A_b A_f) \cdot A_p^*] \tag{6.4}$$

Table 6.3 The relationship between polarized waves and polarization in degenerate four-wave mixing (cubic crystals)[7]

Combi-nation	Polarized wave condition	Four-wave polarization				Polarization component in Equation (6.4)
		f	b	p	c	
1	Linear	↑	↑	→	→	2 C
2	Linear	↑	→	↑	→	A
3	Linear	↑	→	→	↑	A
4	Linear	↑	↑	↑	↑	A, 2 B
5	Circular	R	R	L	L	B−C
6	Circular	R	L	R	L	A, 2 C, B−C
7	Circular	R	L	L	R	A, 2 C, B−C
8	Circular	R	R	R	R	A, B−C

Note: All entries assume that the direction of the polarization vector and the propagation direction are along the crystal's major axis.

f: forward pump wave; b: back pump wave; p: probe wave; c: phase-conjugate wave; →: direction of linear polarization; R and L: right- and left-hand circular polarization

where $A = 3(\chi^{(3)}_{xxyy} + \chi^{(3)}_{xyxy})$, $B = 3(\chi^{(3)}_{xxxx} - \chi^{(3)}_{xxyy} - \chi^{(3)}_{xyxy})$, and $C = 3\chi^{(3)}_{xyyx}$. We can find the polarization relationship between pump, the probe, and phase-conjugate waves from Equation (6.4) as shown in Table 6.3.[7] For example, if we use combination number 1 of polarized waves, the direction of polarization of the probe wave agrees with that of the phase-conjugate wave, as described above. The third-order nonlinearity $\chi^{(3)}_{xyyx}$ contributes in the case where the pump wave is orthogonally polarized. The term $(A_f \cdot A_b)$ in Equation (6.4) contributes only in restricted cases such as two-photon absorption. For isotropic symmetry to be high, Equation (6.4) agrees with Equation (3.13) if we add the stipulation that $\chi^{(3)}_{xxxx} = \chi^{(3)}_{xxyy} + \chi^{(3)}_{xyxy} + \chi^{(3)}_{xyyx}$.

h. Frequency Discrimination

Frequency discrimination employs the difference between the pump and probe frequencies, but it involves non-degenerate four-wave mixing. If we assume the frequency of the two pump waves to be ω_0 and that of the probe wave to be $\omega_0 + \Omega$, the frequency of the phase-conjugate light becomes $\omega_0 - \Omega$ due to the law of conservation of energy. Since spatial separation cannot be used in a nearly-collinear configuration, the frequencies on the frequency axis must be detected separately. If the difference in frequency Ω is very small, a scanning Fabry-Perot interferometer can be used to discriminate between them. This method is being used in semiconductor amplifiers and for nonlinear polarization.

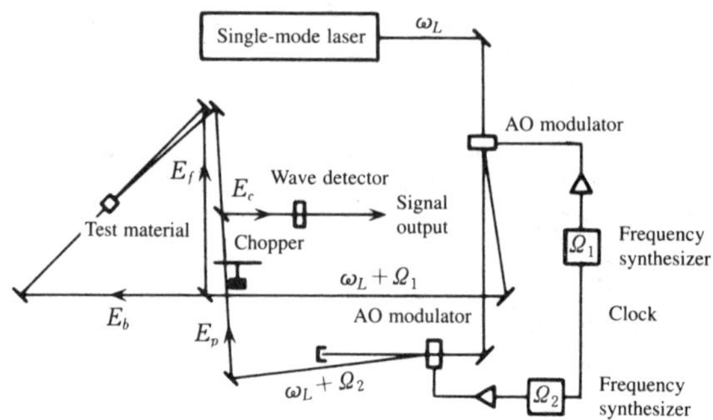

Figure 6.4 Experimental phase-conjugate light generating setup in which the pump and probe wave frequencies are varied by using acousto-optic (AO) modulators. *(After Rand)*[8]

Incidentally, there are methods for producing two light frequencies by varying the frequencies of the pump and probe waves: (i) a technique using two light sources to vary the oscillation frequencies, and (ii) an electro-optic modulator and acousto-optic (AO) modulator configured to use a single light source. Figure 6.4 is an example that uses an AO modulator to shift the frequencies.[8] The light source is a vertical, single-mode Kr ion laser (wavelength: 406.7 nm). The first AO modulator shifts the frequency by only Ω_1; this wave is used as a counterposed pump wave. The frequency is shifted by only Ω_2 by the second AO oscillator and is used as the probe wave. Since an insignificant frequency shift is realized by one AO modulator, two AO modulators are used. The two AO modulators synthesize the basic oscillator's frequency and synchronize the phase. Here $\Omega = \Omega_1 - \Omega_2$. In this example, the probe wave is turned on and off at a low frequency (10 Hz) and detects the phase synchronization of the phase-conjugate wave.

i. Time-Domain Discrimination

Phase synchronization detection is often used to detect weak light to avoid the influences of stray and background light. Since the production of phase-conjugate light involves complex incident waves, a double-chopping method (use of two choppers) is employed to vary the frequency of the pump and probe waves. For example, a probe wave is chopped at 2 kHz and its reference signal and the signal from the phase-conjugate wave detector are input to the first stage of a lock-in amplifier with a time constant of 3 ms. A 108-Hz reference signal from the double-chopped pump wave and the output signal from the first stage are then input to the next stage of the lock-in amplifier with a time constant of 100 ms, allowing the phase-conjugate wave intensity to be measured.[9]

The method described is an extension of so-called phase-synchronization detection and is used when the intensity of the phase-conjugate wave is somewhat weak. The pump and probe waves are derived from a continuous beam by chopping at various time intervals in degenerate four-wave mixing.[4] Phase-conjugate light is produced only when the pump and probe waves are present simultaneously. When the two waves are not temporally superposed, the detected waves are reflected and diffracted background waves emanate from the phase-conjugate wave-generating medium. Therefore, the phase-conjugate wave intensity can be measured by subtracting the value obtained when the pump and probe waves are not superposed from the value obtained when they are superposed.

j. Self-Correlation Method

Another weak-signal detection method makes use of the characteristics of phase-conjugate waves. When phase-conjugate waves are generated by spherical particles in a liquid, light scattering increases due to Brownian motion; this creates background waves. The self-correlation technique can be used to eliminate these background waves. Although the intensity of the phase-conjugate waves can be written as $I_c = \beta I_f I_b I_p$ (I_j: the intensity of each light wave; β: a constant — see Equation (3.30)) when the reflectance is low in counter-propagating degenerate four-wave mixing, in reality background waves are included and can be detected. Accordingly, if only the forward pump wave is double-chopped, for example, the total detected wave intensity becomes

$$I_T(t) = I_{fs}(t) + I_{bs} + I_{ps} + \beta I_f(t) I_b I_p \tag{6.5}$$

where subscript s represents the scattered waves, and the term containing (t) is a function of time. Now, if $< >$ represents time averaging, and if we define the autocorrelation function as

$$g(\tau) = \langle I_T(t) I_T(t+\tau) \rangle \tag{6.6}$$

the wave becomes a triangular chopped wave of period $2\tau_0$ (τ_0: chopper time width). The following relationship incidentally emerges among the measured values[10]

$$\langle I_T \rangle \left[\frac{g(0) - g(\tau_0)}{2} \right]^{1/2} = \langle I_{fs}(t) \rangle + \langle I_c(t) \rangle \tag{6.7}$$

where $\langle I_{fs}(t) \rangle$ is proportional to the beam power P. The phase-conjugate wave intensity can also be found by using data for $\langle I_c(t) \rangle = 0$ because $\langle I_c(t) \rangle$ can be considered proportional to P^3.

The ten techniques described above do not have to be used independently of each other. There are also cases where they are used in combination.

6.3 Four-Wave Mixing Experiments

In this section we introduce the major issues for particular materials associated with various experiments on degenerate four-wave mixing while also showing typical data. The objective in using four-wave mixing to produce phase-conjugate light is not only to increase the reflectance, but also to control various parameters. Applications to materials research that depend on measuring the intensity of the phase-conjugate waves are also being performed. Both characteristics of phase-conjugate light experimentation are laid out in Table 6.4.

(1) Optical Kerr Media
a. Na Metal Vapor

As explained in Section 6.2(2), electron-migration resonance effects are used to increase the nonlinearity of optical Kerr media. We will explain an experimental example here by using Na vapor as the optical Kerr medium.[2] The

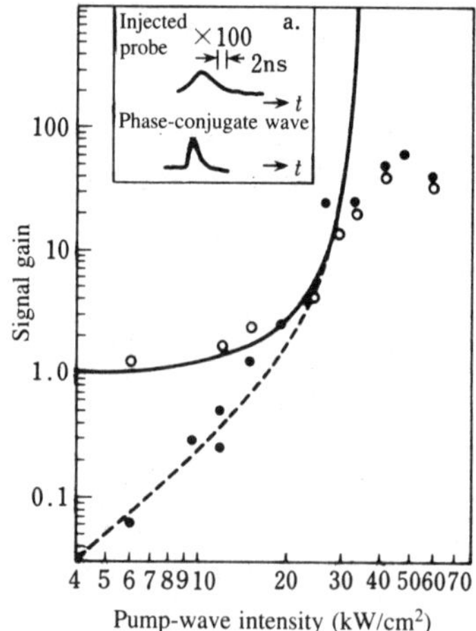

Figure 6.5 Relationship between signal gain and pump-wave intensity for degenerate four-wave mixing in Na vapor. *(After Bloom, Liao and Economou)*[2]

Solid circles: phase-conjugate wave; open circles: transmitted probe wave.
The dashed and solid lines are theoretical curves for optimum fit from Equations (3.28) and (3.29), respectively.
The waveforms in the inset are oscilloscope waveforms for the injected probe wave pulses and the phase-conjugate wave pulses.

Table 6.4 Characteristics of experiments in phase-conjugate light generation for four-wave mixing

Substance name Symbol (unit)	Wave-length (μm)	Laser		Beam power	Nonlinearity $\chi^{(3)}$ (esu)	Reflectance %	Characteristic	Bibliographic reference
		Type	Operating mode					
Rhodamine 6G	0.249	KrF		$25MW/cm^2$	—	2.5	Ultraviolet	11
CS_2 + hexane	0.266	Nd : YAG (FHG)	Pulse	400 μJ	—	0.1	Ultraviolet	12
Diamond	0.407	Kr	Continuous wave	$5W/cm^2$	—	~0.1		8
Eosin	0.514	Ar	Continuous wave	$1W/cm^2$	—	~0.01	Thin-film dye	13
Latex	0.514	Ar	Continuous wave	100 mW	5.8×10^{-8}	~0.47	Particles in liquid	14
Polystyrene	0.514	Ar	Continuous wave	600 mW	1.7×10^{-7}	—	Particles in liquid	10
Polydiacetylene	0.532	Nd : YAG (SHG)	Mode-locked	$1.5 GW/cm^2$	5×10^{-12}	100	In toluene	15
Dyed glass filter	0.532	Nd : YAG (SHG)	Q-switched	$1 MW/cm^2$	1.3×10^{-8}	~10	Semiconductor particles	16
Organic azo dye	0.532	Nd : YAG (SHG)	Q-switched	10 mJ	1.0×10^{-10}	80	In solution	17
Hg atmosphere	0.546	Dye		50 mW	—	~0.1		18
Na atmosphere	0.589	Dye	Continuous wave	1.2 W	—	150		19
Na atmosphere	0.589	Dye	Pulse	7.5 kW	2.9×10^{-10}	~10^4		2
DODCI	0.605	Dye	Mode-locked	50 μJ	—	50	Dye	20
CS_2	0.694	Ruby	Q-switched	—	1.6×10^{-12}	250		6
MBBA*	0.694	Ruby	Q-switched	10 mJ	—	230	Circular polarized excitation	21
GaAs	0.694	Ruby	Q-switched	$0.3 J/cm^2$	—	5		22
Rb atmosphere	0.795	Dye		$2 kW/cm^2$	—	—		23
Si	1.06	Nd : YAG	Q-switched	200 mJ	8×10^{-8}	150		24, 25
Ge	10.6	CO_2	TEA	$100 MW/cm^2$	2.5×10^{-11}	800		26
KCL : ReO$_4$	10.6	CO_2	Mode-locked	>$20 MW/cm^2$	—	~1	Resonance absorption	27
HgTe	10.6	CO_2	Q-switched	$1 MW/cm^2$	1.6×10^{-4}	—		28
HgCdTe	10.6	CO_2	TEA	$750 kW/cm^2$	5.4×10^{-6}	10		29
SF_6	10.6	CO_2	TEA	$1 MW/cm^2$	—	7	Resonance absorption	30
InSb	10.6	CO_2	Q-switched	1 kW	2×10^{-7}	—	(77k)	31
$BaTiO_3$	0.515	Ar	Continuous wave	5 mW	—	2000	Photorefractive effect	32
$Bi_{12}SiO_{20}$	0.568	Kr	Continuous wave	$140 mW/cm^2$	—	270	Moving diffraction grating	33
GaAs	1.06	Nd : YAG	Continuous wave	40 mW	—	500	Moving diffraction grating	34

configuration is retroreflective. The light source is an N_2-excited dye laser with a 6-ns pulse width. A wavelength 1.25 cm^{-1} below the Na D$_1$ resonance was established in order to avoid strong absorption effects. The beam from the optical source was parallel and had a 3-mm diameter; peak power was approximately 7.5 kW. The beam was injected into a sealed cell (5 cm long) containing an Na vapor concentration of 3×10^{14} cm^{-3} and a He buffer at 5 Torr. The retroreflective configuration had an approximately 0.5° angle of intersection between the pump and probe waves. The phase-conjugate waves produced were detected by a fast optical detector after reflection by a beam splitter. Figure 6.5 shows the measured gains. Maximum reflectance was about 100. The solid circles represent reflected waves; the open circles are the transmitted waves. The broken and solid lines represent the theoretical curves when absorption and pump light attenuation are neglected [$R = \tan^2(\varkappa L)$, $T = \sec^2(\varkappa L)$: Equations (3.28) and (3.29)]. If the length L is a measured 5 cm, we obtain $n_2 = 1.1 \times 10^{-8}$ esu (1.2×10^{-17} m^2/V^2). Although the theoretical and measured results are in good agreement up to pump wave intensities of 30 kW/cm^2, a shift is produced due to saturation effects at higher intensities.

A value equivalent to $n = 1$ is obtained when the beam source is set to a wavelength halfway between the Na D$_1$ and D$_2$ resonances. Scattering as the beam propagates through the medium is replaced by the generation of phase-conjugate light.[35]

b. CS$_2$

An experimental setup using a liquid as the nonlinear medium is shown in Figure 6.6.[6]

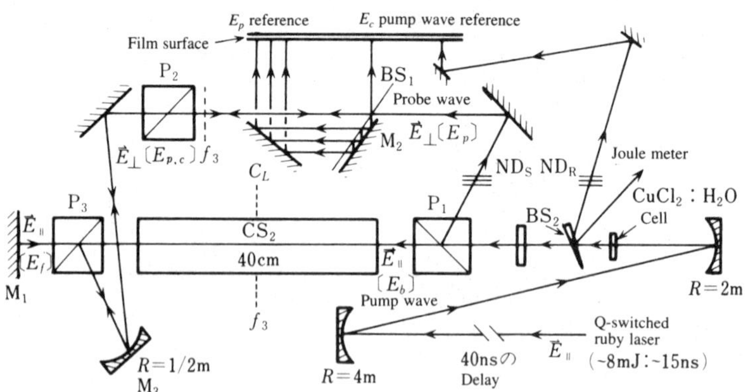

Figure 6.6 Experiment with phase-conjugate wave generation in CS$_2$ (using polarization discrimination). *(After Pepper, Fekete and Yariv)*[6]

E_f, E_b: pump waves (polarized in the plane of the page)
E_p: probe wave (orthogonally polarized in the plane of the page)
E_c: phase-conjugate wave (orthogonally polarized in the plane of the page)

Vertical- and horizontal-mode Q-switched ruby lasers (wavelength: 0.694 μm, pulse width: 15 ns, pulse energy: $7-13$ mJ) are used as the light source. A pair of concave mirrors supply a 40-ns delay until the waves emanating from the light source reach the CS_2 cell; the waves coming from the cell do not coincide with the outgoing pulses. A $CuCl_2:H_2O$ cell is used to vary the concentration in order to control the transmitted wave power. The waves transmitted by polarized beam splitter P_1 become the pump waves and are injected into the nonlinear CS_2 medium. This is a waveguide sealed in a glass tube whose length is 40 cm and whose inside diameter is 2 cm. The pump waves passing through the CS_2 are retroreflected by mirror M_1. After the waves reflected by polarized beam splitter P_1 are reflected, they become probe waves. Polarized beam splitters P_1, P_2 and P_3 are configured so that the pump wave is parallel to the glass tube and the probe and phase-conjugate waves are perpendicular to it. The phase-conjugate waves are recorded on ASA 3000 high-speed oscilloscope film. The reflectance R is almost 1 when the pump wave energy is 11 mJ. If we look at the nonlinearity by curve-fitting $R = \tan^2(\xi I_f)$ (ξ: constant of proportionality; I_f: pump-wave energy), we obtain $\chi^{(3)}_{yxxy} = 1.64 \times 10^{-12}$ esu (2.3×10^{-20} m^2/V^2).

The initial experiments with degenerate four-wave mixing were performed by using CS_2 as the optical Kerr medium with a frequency-doubled Q-switched Nd:YAG laser as the light source.[36] CS_2 is currently the standard material being used.

CS_2 is transparent into the visible and infrared spectra; however, due to high electron absorption it is opaque at under 360 nm at thicknesses of 220 and 310 nm. Accordingly, CS_2 has a phase-conjugate reflectance of $R = 0.1\%$ in the ultraviolet range when it is mixed with transparent hexane (60 vol % of CS_2).[12]

(2) Alkali Halides

Alkali halides can be used to easily explain the experimental results obtained by the saturation absorption mode.[27] The test material used was $KCl:ReO_4$ with lengths of 1 and 2 cm. The light source was a mode-locked CO_2 laser (wavelength: 10.67 μm, pulse width: 2 ns); a counterpropagating-pump configuration was used. The detector was Ge:Au, and could measure up to 10 nJ. Figure 6.7 shows the calculated reflectance as a function of pump intensity when both pump waves I_f and I_b have the same intensity. The maximum pump wave intensity is 20 MW/cm^2, more than ten times that of the saturated wave intensity $I_{sat} \approx 1.45$ MW/cm^2. The saturation characteristics are clearly expressed by this figure, in which transition between oscillation levels depending on the ReO_4^- ions in the KCl grating is in resonance with the CO_2 P(28) transition. The solid curve represents the theoretical values from Equation (4.21) assuming a 2-level atomic system as the saturation-absorption model. Although a slightly larger

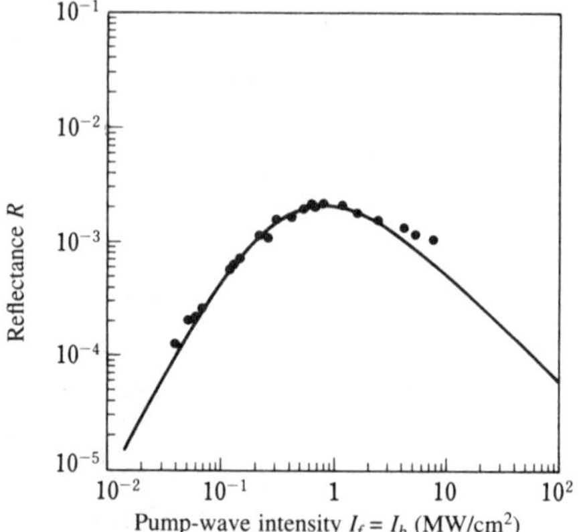

Figure 6.7 Reflectance saturation characteristics in KCl:ReO$_4$.
(After Watkins, Figueira and Thomas)[27]

Wavelength: 10.67 μm; pump wave intensity $I_f = I_b$;
       length of test material $L = 1$ cm; $\alpha_0 L = 0.47$;
       α_0: small-signal absorption coefficient.
Solid curve: The theoretical value obtained from Equation (4.21)
       with $\alpha_0 L = 0.52$.

absorption coefficient α_0 was used in calculating the theoretical values in order to make the experimental values fit well, they do fit the trend well. The absolute values of the experimental and theoretical values agreed to within 50% accuracy.

The same kind of measurements were then made for high-absorptivity test materials ($\alpha_0 L = 3.0$). Although the experimental and theoretical curves were in fairly good agreement as to trend, the absolute value of the reflectance was five times higher than the theoretical value. The pump-wave intensity was attenuated more than 80% in the experiment, but in theory this attenuation was not considered.

(3) Semiconductors

To begin with, we will explain typical examples of the measurement of phase-conjugate reflectance R for semiconductors.[24] Specimens of thickness L were cut from identical wafers of three types of p-Si. The carrier density was less than 10^{20} m^{-3}. Since the small-signal absorption coefficient was $\alpha_0 \sim 10$ cm^{-1}, the thickness of the test material was approximately 1 mm because $\alpha_0 L \approx 1$. The optical source was a single-horizontal-mode Q-switched Nd:YAG laser (wavelength: 1.06 μm; pulse width: 15 ns; output: 200 mJ; coherence length: approximately 2 m); a counterpropagating-pump configuration was used. When the forward pump wave E_f and the probe wave E_p were orthogonally polarized, phase-conjugate waves were not quite produced. A large R was obtained when the E_f and E_p polarizations coincided. When the angle θ formed by E_f and E_p was decreased, R increased monotonically, but for convenience the angle for the optical configuration was maintained at $\theta \approx 2°$.

The relationship between phase-conjugate wave reflectance and pump wave intensity in the experiment described above is shown in a logarithmic plot in Figure 6.8. The R for low pump-wave intensity has a slope of 2 for any material thickness. The characteristics expected from Equation (3.30) are shown as well. The nonlinearity $\chi^{(3)}$ is estimated to be 8×10^{-8} esu (1.1×10^{-15} m²/V²). However, a saturation R due to an increase in free electrons accompanied the increase in pump intensity. Thick material, in particular, saturates at a lower pump-wave intensity. This saturation was explained in the discussion of saturable absorbing media in Section 4.1. The maximum reflectance obtained was about 150%.

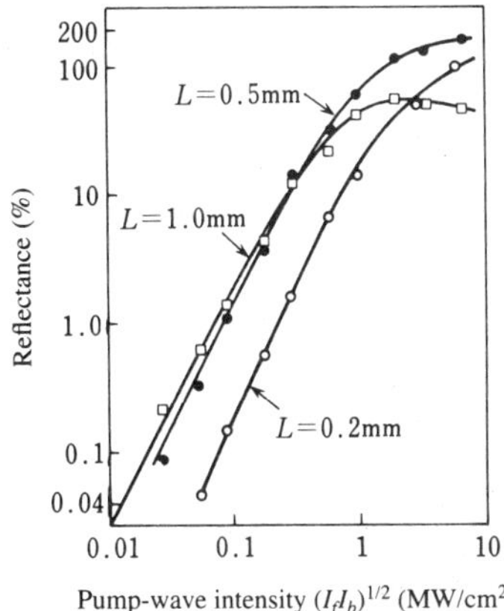

Figure 6.8 Reflectance versus pump-wave intensity for Si. *(After Jain, Klein and Lind)*[24]

Wavelength: 1.06 μm; pump wave intensity $I_f \approx I_b$; probe wave intensity: $I_p = 0.02$ MW/cm².

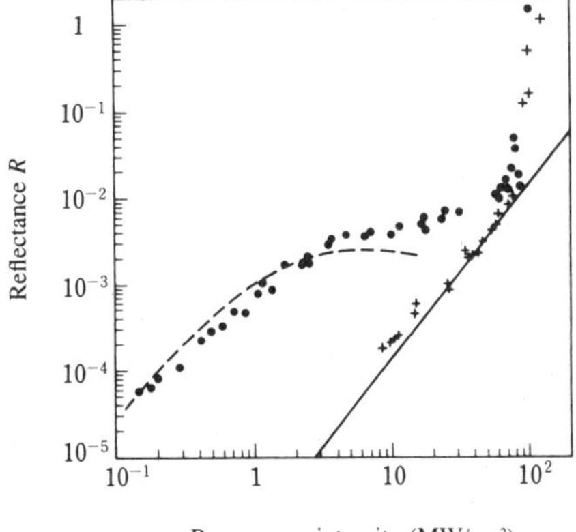

Figure 6.9 Reflectance versus pump-wave intensity for Ge. *(After Watkins, Phipps, Jr. and Thomas)*[26]

Black circles: p-Ge; plus signs: optical-quality Ge; depth: 3 mm; wavelength: 10.6 μm.

Phase-conjugate reflectance saturates as the pump wave intensifies. The characteristic of decreased reflectance above a certain wave intensity has been observed in Si (1.06 μm),[25,37] HgCdTe (10.6 μm),[29] and CdS (0.532 μm),[16] among other compounds.

When a complex mechanism acts as the source of nonlinear polarization, a separate mechanism begins to operate in regard to increases in pump-wave intensity, resulting in a sudden increase in phase-conjugate reflectance. An example of this is shown in Figure 6.9.[26] The test materials were optical crystals and p-Ge with a thickness of $L = 3$ mm. The light source was a TEA CO_2 laser (wavelength: 10.6 μm; pulse width: 1.5 ns), and the setup was a counter-propagating-pump configuration. The angle formed by the pump and probe waves was approximately 4° in the specimen. The generation of a free-electron plasma dramatically increased the reflectance when the pump intensity I_f was greater than 80 MW/cm^2; the reflectance was proportional to the eleventh power of I_f for $80 < I_f < 120$ MW/cm^2. The production of a free-electron plasma has been confirmed due to the reinforcement of the reflectance only when optical conductance increases. When the hole density is $N_h = 2.5 \times 10^{21}$ m^{-3} and $I_f/I_{sat} = 125$ (I_{sat}: saturation intensity), the theoretical refractive index is $\delta n = -6.4 \times 10^{-4}$ and the theoretical reflectance when absorption by the medium is neglected is $\tan^2(k\delta nL)$ (k: wave number in free space). This is shown by the solid line in the figure. The p-Ge mechanism below the plasma-generation threshold can be explained by resonance effects based on absorption by holes. A dramatic increase in the reflectance by the same type of increase in wave intensity has been observed for HgCdTe.[38]

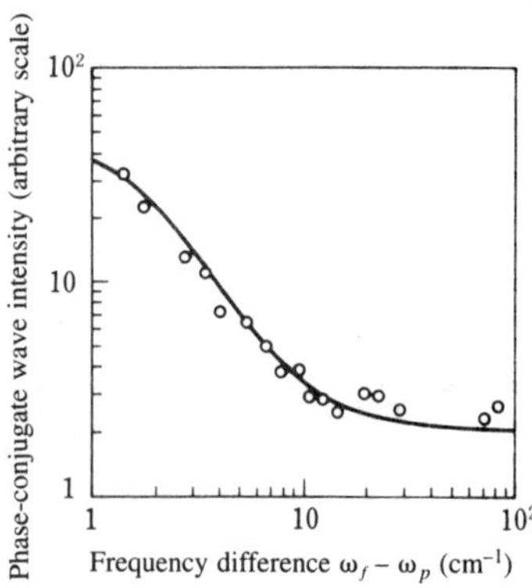

Figure 6.10 Phase-conjugate wave intensity versus detuning for nearly-degenerate four-wave mixing. *(After Yuen and Wolff)*[31]

Test material: n-InSb; light source: CO$_2$ laser (10.6 μm band); temperature: 77 K.
The vertical axis is in proportion to $\chi^{(3)}$ and the theoretical value is found from $\chi^{(3)} = \chi_v^{(3)} + \chi_f^{(3)}$.
$\chi_v^{(3)}$: contribution of bound electrons; $\chi_f^{(3)}$: effect of temperature fluctuations; ω_f: pump-wave frequency; ω_p: probe-wave frequency.

The relationship between phase-conjugate wave reflectance and detuning as a result of nearly-degenerate four-wave mixing is shown in Figure 6.10.[31] The test material was n–InSb (doped with Te) having a carrier density of 2×10^{23} m^{-3}, a thickness of 150 μm, and a measured temperature of 77 K. The light source was a pair of synchronized Q-switched CO$_2$ lasers (10.6 μm band). Reflectance measurement was performed as the difference between the probe-wave frequency ω_p and the pump-wave frequency ω_f was varied. We see from the diagram that if detuning between ω_p and ω_f decreases, there is a striking increase in the reflectance. The solid curve represents the theoretical values which are found from $\chi^{(3)} = \chi_v^{(3)} + \chi_f^{(3)}$, where $\chi_v^{(3)} = 1.3 \times 10^{-9}$ esu is contributed by bound electrons (Equation (7.17)), $\chi_f^{(3)}$ is found from Equation (7.20) with $\tau_m = 1.4$ ps, $\tau_{th} = 3$ ps, and $\chi_{wp}^{(3)} = 1.3 \times 10^{-8}$ esu. $\chi^{(3)} = 2 \times 10^{-7}$ esu (2.8×10^{-15} m^2/V^2) is obtained from the peak-value data; the electron thermal energy relaxation τ_{th} lies in the range of 2 – 4 ps. This method yields positive nonlinear polarization.

Similar detuning dependence data has been obtained for Si[39] and HgTe.[28] For HgTe in particular, when the wavelength is 10.6 μm a value as large as $\chi^{(3)} = 1.6 \times 10^{-4}$ esu is obtained at room temperature, with an absorption coefficient as high as $\alpha = 3.4 \times 10^3$ cm^{-1}.

(4) Non-Crystalline Substances

One objective of commercially-available long-wavelength band-pass filters containing semiconductor particles precipitated in glass is a steep cutoff characteristic. For example, filters using CdS_xSe_{1-x} have a cutoff of approximately 600 nm. Characteristics different from those of implanted semiconductors are exhibited when these dyed-glass filters are used as phase-conjugate wave producing media.

A frequency-doubled horizontal single-mode Q-switched Nd:YAG laser (wavelength: 0.532 μm; pulse width: 10 ns) was used as the light source for investigating characteristics of dyed-glass filters.[16] A counterpropagating-pump configuration was employed, with the angle made by the pump and probe

	Wave polarization condition	Relative phase-conjugate wave intensity	
		$CdS_{0.9}Se_{0.1}$ glass	CdS crystal
a		1	1
b	Long-period diffraction grating due to E_f and E_p	0.35 ($\pm$0.07)	1
c	Short-period diffraction grating due to E_b and E_p	0.45 ($\pm$0.08)	0
d	Temporal diffraction grating due to two-photon absorption	0	0

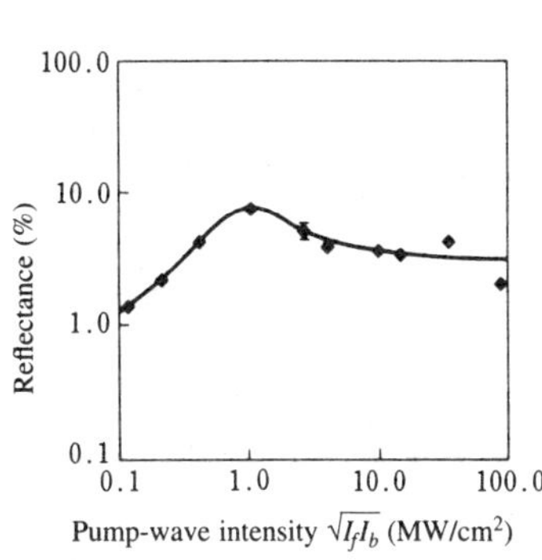

Figure 6.11 Relationship between phase-conjugate wave intensity and pump-wave frequency for $CdS_{0.9}Se_{0.1}$ glass. *(After Jain and Lind)*[16]

Thickness of test material: 2.0 mm; wavelength: $\lambda = 0.53$ μm.

Figure 6.12 Relationship between phase-conjugate wave intensity and polarization conditions at wave injection for $CdS_{0.9}Se_{0.1}$ glass and CdS crystal. *(After Jain and Lind)*[16]

Injected wave intensity: $I_f = I_b \cong 50$ MW/cm^2; $\lambda = 0.532$ μm.

waves set at approximately 2°. Figure 6.11 shows phase-conjugate reflectance versus pump-wave intensity when a dyed-glass filter was used (Corning No. 3484/CS 3–68, thickness: 2 mm). When the pump-wave intensity was low (~ 0.1 MW / cm^2), the reflectance was higher with the dyed-glass filter than it was with a CdS crystal. Saturation characteristics were exhibited as the wave intensity increased. The nonlinearity $\chi^{(3)}$ of this dyed-glass filter is of the order of 1.3×10^{-8} esu (1.8×10^{-16} m^2/V^2) — 10 times greater than that for a CdS crystal.

Figure 6.12 shows relative values of dyed-glass filter reflectance as pump- and probe-wave polarizations vary.[16] For the sake of comparison, we also show measured results for CdS crystals using the same setup. A characteristic of dyed-glass filters is that the values are almost the same when the diffraction grating has a long period ($\Lambda_{fp} = 3$ μm) as when it has a short period ($\Lambda_{bp} = 0.1$ μm). It is believed that electron and hole carrier distributions serve the purpose of forming small-period diffraction gratings because particles of 0.1 μm or smaller exist in the dyed-glass filter. On the other hand, only a long-period diffraction grating and not a short-period grating makes a contribution in the case of CdS. This is because the depth of a short-period diffraction grating is reduced to a diffusive contribution in the picosecond range. In this connection, the relative contribution to $\chi^{(3)}$ for both diffraction gratings is proportional to $(\Lambda_{fp}/\Lambda_{bp})^2$.

(5) Photorefractive Media

As explained in Section 3.4, experiments with photorefractive media make use of highly nonlinear substances, with milliwatt continuous-wave lasers as light sources for phase-conjugate light production. We next explain some typical examples.

(a) BaTiO$_3$

As a photorefractive medium, BaTiO$_3$ has an electro-optic coefficient as high as $r_{42} = 1640$ pm/V. Since it is available in the form of relatively large high-quality optical crystals, BaTiO$_3$ can be easily used for the production of phase-conjugate light. The crystal has a 4*mm* structure and a low degree of symmetry and, in addition, since BaTiO$_3$ exhibits birefringence, the angle between the beams must be rigorously established when the optical system is set up.

Figure 6.13(a) shows the beam angles for degenerate four-wave mixing. θ_1 is the angle formed within the crystal by its c-axis and the forward pump wave E_f. θ_2 is the angle formed by the c-axis and the probe wave E_p. Consider reflection from the diffraction grating formed by E_f and E_p on which the forward pump wave E_b is Bragg-diffracted. When all four waves are p polarized, the reflectance for the ordinary back pump wave can be approximated by[32]

$$R_{ord} = \left| \frac{\omega L E_s \xi}{4c} n_o{}^3 r_{13} \cos \theta_G \right|^2 \tag{6.8}$$

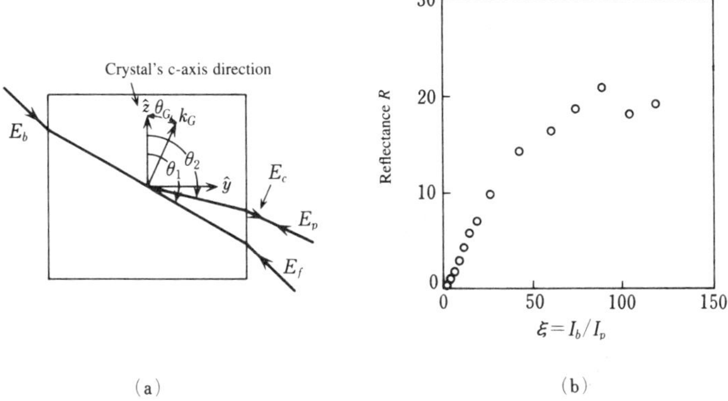

Figure 6.13 Phase-conjugate wave generation using BaTiO$_3$. *(After Feinberg and Hellwarth)*[32]

(a) Beam configurations
(b) Reflectance measurement example (when all waves are extraordinary rays)
$I_p = I_f/4$; $\theta_1 = 106°$; $\theta_2 = 114°$; $\theta_G = 20°$.
Light source: Ar ion laser (514.5 nm)

and the reflectance for the extraordinary back pump wave can be approximated by

$$R_{ext} = \left| \frac{\omega L E_s \xi}{4 c n_3} \cos \theta_G \left(n_e{}^4 r_{33} \sin \theta_1 \sin \theta_2 + 2 n_e{}^2 n_o{}^2 r_{42} \sin^2 \theta_G \right. \right.$$
$$\left. \left. + n_o{}^4 r_{13} \cos \theta_1 \cos \theta_2 \right) \right|^2 \tag{6.9}$$

where $\xi \equiv I_b/I_p$, $n_o = 2.49$ ($n_e = 2.42$) is the refractive index for the ordinary (extraordinary) ray, n_3 is the refractive index for the read ray, E_s is the field magnitude, θ_G is the angle formed by the diffraction grating vector and the c-axis, r_{ij} is the electro-optic coefficient, L is the effective interaction length and c is the speed of light in vacuo. Since r_{42} when compared to other r_{ij} is usually large for BaTiO$_3$, the reflectance for extraordinary rays is generally high.

Figure 6.13(b) shows an example of measured values of phase-conjugate wave reflectance R for BaTiO$_3$ where all waves are taken to be extraordinary.[32] Based on Equation (6.9) the beam angles are $\theta_1 = 106°$, $\theta_2 = 114°$, and $\theta_G = 22°$ for high values of R. The light source is an Ar ion laser ($\lambda = 514.5$ nm). The size of the crystal is $2.2 \times 2.8 \times 4.2$ mm^3 ($L = 0.4$ cm). The angle between the c-axis and the grating vector was increased and the crystal was immersed in a refractive liquid (refractive index: 1.51) to increase the separation between the pump and probe waves. High-gain amplification characteristics of BaTiO$_3$ were clearly observed from the experiment. The experimental values agree to one decimal place with the values calculated from Equation (6.9). Equations (6.8) and (6.9) were solved with coupling between E_p and E_b only. In reality, however, there is also coupling between E_p and E_f, producing a disparity between the resulting experimental and calculated values for extraordinary rays. Nevertheless, Equations (6.8) and (6.9) are useful for calculating absolute values of R for all beam angles.

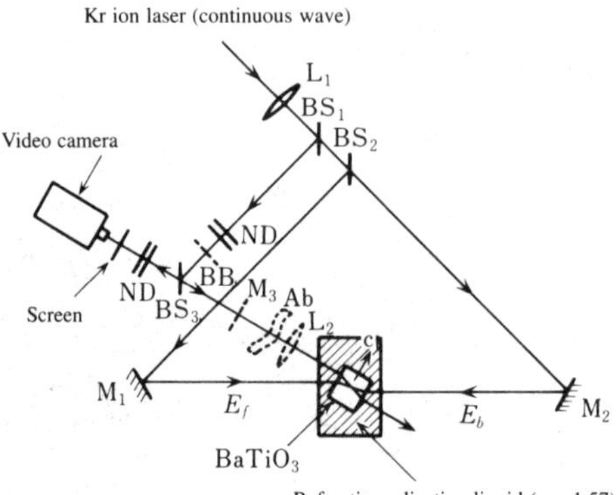

Figure 6.14 Schematic diagram of a phase-conjugate wave generating experiment using BaTiO₃. *(After Jain and Dunning)*[40]

BS: beam splitter; M: reflecting mirror.
The arrow (c) represents the direction of the crystal's c-axis.

Figure 6.14 shows an example of a setup for producing phase-conjugate waves by using a BaTiO₃ crystal.[40] The light source was a horizontal, single-mode, continuous-wave Kr ion laser (wavelength: 0.6471 μm; output: approximately 100 mW) with an opto-isolator inserted between the light source and lens L_1. Beam splitter BS_1 had 2% reflectance. The transmitted wave was the pump and the reflected wave was the probe in this counterpropagating configuration. All light waves were p polarized. The crystal was immersed in a refractive liquid (refractive index: 1.57) before the experiment but after it was poled (poling: approximately 15°C from the Curie point and up to 145°C; electric field: approximately 1.11 kV/mm impressed for 2 to 3 hours). Reflectance in excess of 2 was observed in a pump-wave power range of 5 – 100 mW at various wave intensity ratios ($I_b/I_f = 1 - 4$, $I_f/I_p = 1 - 10$).

b. BaTiO₃ in a Self-Excitation Arrangement

As previously mentioned, BaTiO₃ shows reflectance greater than unity and can amplify. This means that it has high internal gain and suggests the possibility of self-excitation. Let us consider the Fabry-Perot etalon in Figure 6.1(d). If the probe wave E_p has high intensity here, a diffraction grating is formed in the medium by the E_p and scattered waves. If the internal gain of a nonlinear medium is higher than the reflective or absorptive losses, E_p can be Bragg-diffracted by this diffraction grating, resulting in a new wave that oscillates within the resonator. In addition, the new wave is diffracted due to the diffraction grating that is formed by the E_p and scattered waves, resulting in a phase-conjugate wave E_c emerging in the direction opposite that of E_p. When an He-Ne laser (output: approximately 1 mW) is used as the light source, approximately 10% reflectance is obtained.[41]

The setup resembles the ring resonator in Figure 6.1 (e). The mechanism that produces phase-conjugate light is similar to the Fabry-Perot etalon. The probe itself is Bragg diffracted in these configurations, slowing the response time until the diffraction grating is formed. This is approximately 100 seconds for the Fabry-Perot resonator and about 8 seconds for the ring resonator.[41]

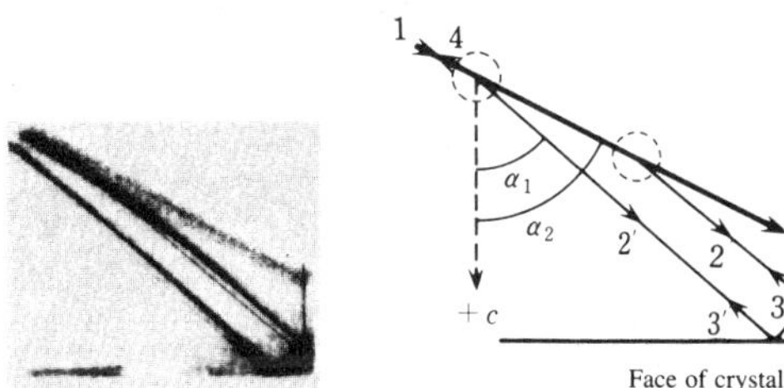

Figure 6.15 Phase-conjugate wave generation with an internally reflective crystal using $BaTiO_3$. *(After Feinberg)*[42]

The optical resonator described can be modified to take advantage of the crystal's internal reflectance, so that phase-conjugate light can be produced without the use of external mirrors.[42] This internal-reflector arrangement was shown in Figure 6.1(f); an example is shown in Figure 6.15. If the probe wave is injected toward the edge of the crystal, the scattered wave (2) is reflected twice by the high-gain $BaTiO_3$ medium and becomes 3′. This beam intersects beam 1, produces beam 2′ at this location, and propagates in the direction opposite to beam 2. Phase-conjugate wave 4 is produced when these waves act as two pump waves and mix with the injected probe by four-wave mixing (shown in the broken circle). An Ar ion laser (514.5 nm) and an Ar-excited dye laser (dye: rhodamine 6G; variable wavelength range: 50 nm) was used. A reflectance of approximately 30% was obtained with a crystal size of $5.6 \times 5.4 \times 4.3$ mm^3. The power-density threshold value for self-excitation in this case was 0.027 W/cm^2 (beam power: 1.7 mW; beam diameter: 2.8 mm).

In addition to $BaTiO_3$, self-excitation has also been attempted with CS_2[43] and SBN.[44]

c. Moving Diffraction Grating

$Bi_{12}SiO_{20}$ (BSO) is a high-quality photorefractive medium. Although its electro-optic coefficient $r_{41} = 5$ pm/V is two to three orders of magnitude smaller than that of $BaTiO_3$, its cubic (43m) crystalline structure makes it easy to set up in optical systems. And because the reflectance R of BSO is relatively low (of the order of 0.1%),[45] we can use a moving diffraction grating (described below) as a means for increasing the reflectance.

As mentioned in Section 3.3, the phase difference ψ between the wave-intensity distribution and the refractive-index distribution exerts a great influence on the reflectance in degenerate four-wave mixing based on the photorefractive effect. In BSO, light-excited carriers are accelerated efficiently by an externally-impressed electric field, making use of a moving diffraction grating. A moving-diffraction-grating experiment has been performed with BSO crystals using a counterpropagating setup.[33] The light source was a continuous-wave, single-mode Kr ion laser oscillating at 568 nm. When the pump waves (wave intensities: I_f and I_b) and the probe wave I_p were injected into the crystal — as shown in Figure 6.16 — one of the pump waves (I_f) was injected into the crystal by reflection from a mirror. A piezoelement was attached to the mirror to create sawtooth oscillation with motion in one direction at a constant speed. The frequency of the generated phase-conjugate wave shifted because the I_f frequency shifted slightly

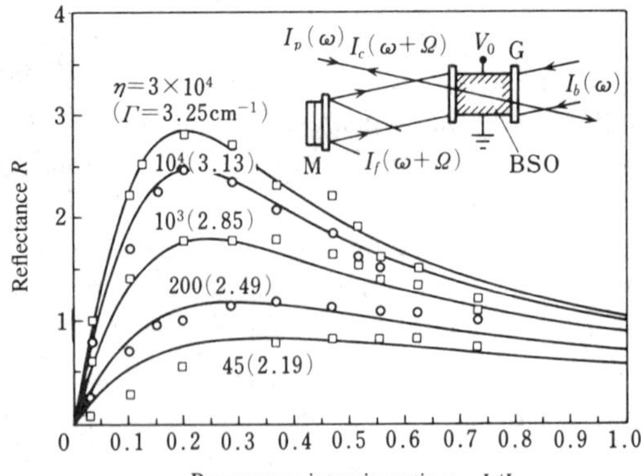

Figure 6.16 Reflectance versus pump wave intensity ratio in BSO using a moving diffraction grating. *(After Rajbenbach, Huignard and Refrégier)*[33]

$\eta = I_f/I_p$; Γ: gain coefficient.
Interference fringe spacing: $\Lambda = 23$ μm; externally impressed electric field: $E_0 = 10$ kV/cm.
I_f, I_b: pump waves; I_p: probe wave; I_c: phase-conjugate wave; M: piezo mirror; G: glass plate with anti-reflection coating.

(Ω) from the fundamental frequency due to the Doppler effect caused by the mirror movement. Consequently, we obtain nearly-degenerate four-wave mixing.

A measured example of the reflectance in the above-mentioned experiment is shown in Figure 6.16.[33] Assume that $I_f = 80$ mW/cm^2, $E_0 = 10$ kV/cm (electrode separation: 5 mm), and the length of the crystal is 10 mm. Assume further the pump-wave intensity ratio $r = I_b / I_f$ and the probe-wave intensity ratio $\eta = I_f/I_p$ as parameters. The velocity of interference-fringe movement is optimized for each measurement point. The maximum reflectance for each η can be obtained from the nonsymmetric pump-wave intensity ratio as explained by the results in Section 3.3(4). When the interference fringe spacing $\Lambda = 23$ μm, $\eta = 3 \times 10^4$, and $r = 0.2$, we obtain $R = 2.7$ for the maximum reflectance. The frequency shift under the experimental conditions described above is of the order of $\Omega = 10$ Hz. The solid curve represents the best fit as derived from Equation (3.74) with $\gamma = (\Gamma/2)(1 - i \cot \psi)$ (Γ: gain coefficient).

d. Fast Response in Photorefractive Media

Although a photorefractive medium — of which BaTiO$_3$ and BSO are representative — may have apparent high nonlinearity, it has the disadvantage of slow response speed. Accordingly, research has focused on achieving (i) fast response in these types of media and (ii) fast response in compound semiconductors. Although GaAs, InP and CdTe are being studied in regard to (ii), they are used in infrared applications, which are different from those of the transparent (visible) regions of oxides. Transient response of photorefractive media for short pulses will be covered in Section 6.5.

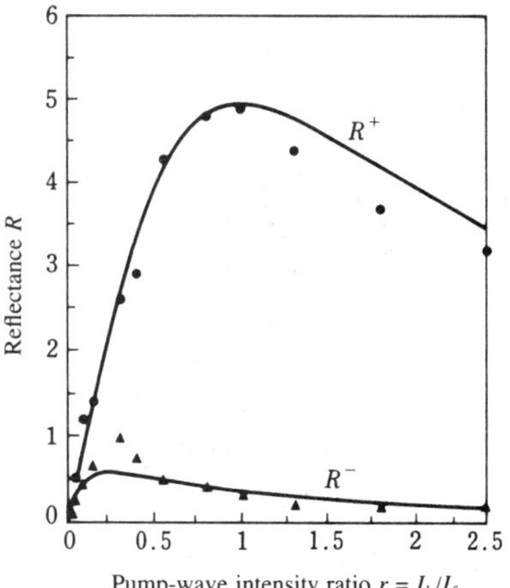

Figure 6.17 Reflectance versus pump-wave intensity for GaAs:Cr. *(After Rajbenbach, Imbert, Huignard and Mallick)*[34]

R^-: Reflectance when the pump- and probe-wave polarizations agree;

R^+: Reflectance when the forward pump- and probe-wave polarizations agree and the back pump wave has orthogonal polarization.

Externally impressed voltage: 4 kV; interference fringe spacing: $\Lambda \cong 20$ μm; $I_f \cong 20$ mW/cm^2; the solid curve represents theoretical values for gain factor $\Gamma = 7.7$ cm^{-1}.

GaAs:Cr is used as a medium in an optical system similar to that in Figure 6.16 where phase-conjugate light is produced by nearly-degenerate four-wave mixing.[34] The light source is a diode-excited, continuous-wave Nd:YAG laser (wavelength: 1.06 μm; optical output: 40 mW), the interaction length of the medium is 3.6 mm, and the electrode separation is 4.7 mm. As explained in Section 3.3(5), the polarization of each light wave was established to allow positive or negative feedback. Figure 6.17 shows the relationship between reflectance and the pump-wave intensity ratio r. The case where the directions of polarization of the probe and the two pump waves agree (negative feedback) is shown by R^-; R^+ indicates that the polarization of the forward pump wave and probe wave coincides with one side of the major axis of the refraction ellipse in the crystal and that it is orthogonal to the back-pump-wave polarization (positive feedback). When positive feedback is used, a maximum reflectance of $R^+ = 5$ is obtained when $r = 1$. The solid curve represents a gain coefficient Γ of 7.7 cm^{-1}, and is the theoretical curve found with $\gamma = i\,(\Gamma/2)\,\exp(-i\psi)$ in Equations (3.74) and (3.77). The experimental and theoretical values agree comparatively well. Both theoretical equations are also good approximations for nearly-degenerate four-wave mixing with a moving diffraction grating.

(6) Organic Compounds
a. Macromolecules

An experiment was performed with poly-4-BCMU polydiacetylene (see Section 6.5(1)) as a macromolecular thin film using a mode-synchronized YAG-excited dye laser (wavelength: 605 nm) as the light source in a counterpropagating configuration.[46] The test material was a 4-μm-thick film made from 4-BCMU dissolved in a chloroform solution. When an optical pulse having a width of 500 fs was injected, a phase-conjugate wave with a pulse width of approximately 700 fs was obtained. A phase change occurred if the temperature was high (in the neighborhood of 370 K). At room temperature, $\chi^{(3)} \approx 4 \times 10^{-10}$ esu (5.6 $\times$ 10^{-18} m^2/V^2; wavelength: 585 – 605 nm), but after the phase change, $\chi^{(3)} \approx 2.5 \times 10^{-11}$ esu. In contrast to the case of macromolecular single crystals, the value of $\chi^{(3)}$ was nearly isotropic in the thin film.

b. Organic Dyes

We next explain an example of organic dyes used as saturable absorbers. Although the dyes tested were rhodamine 6G, DODCI, eosin and erythrosin, among others, we will take up the example of DODCI here.[20] The light source was a flash-lamp-excited, passive-mode, synchronous dye laser (wavelength: 605 nm; pulse width: 5 ps; pulse energy: approximately 50 μJ) used in a retro-reflective configuration. Although ethanol, methanol and glycerol were used as

DODCI solvents, the experimental results did not depend on them. The dye cells were the same as those used for passive-mode synchronization, and produced reflected waves via a 100%-reflective mirror in the cell. The thickness of the dye cell was 0.02 cm. The beam diameter was approximately 1 mm and the peak pump-wave intensity was approximately 500 MW/cm^2.

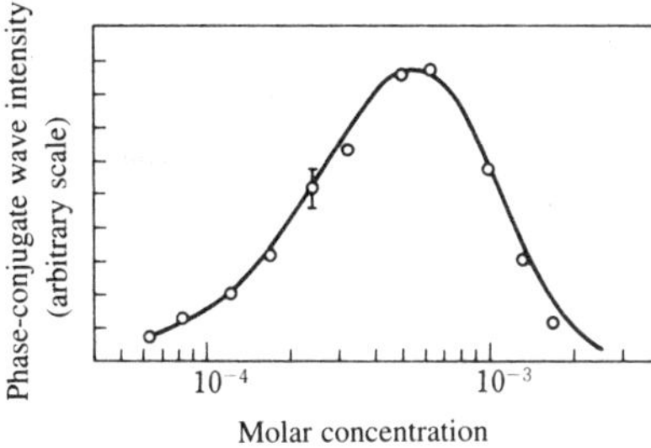

Figure 6.18 Relationship between molar concentration of DODCI organic dye and phase-conjugate wave intensity. *(After Tocho, Sibbett and Bradley)*[20]

Pump-wave intensity: I_f = 440 MW/cm^2; probe-wave intensity: I_p = 36 MW/cm^2.
The solid curve is the theoretical value from Equation (3.37).
Extinction coefficient: $\eta = 5 \times 10^4$/(mole • cm); L = 0.02 cm.

Figure 6.18 shows the relationship between DODCI molar concentration and phase-conjugate wave intensity I_c; there is an optimum level of concentration.[20] Below 5.5×10^{-4} molar, there is a decrease in the phase-conjugate wave intensity due to the low coupling coefficient. On the other hand, absorption increases above this level of concentration, and the two pump-wave intensities become asymmetric due to the retroreflective configuration, reducing the efficiency. The solid curve in the figure is the fitted curve

$$I_c \propto I_f^2 [1 - \exp(-\alpha_0 L)]^2 \exp(-\alpha_0 L) \tag{6.10}$$

This is a modified theoretical curve based on the optical absorption results of Equation (3.37). Here I_f is the pump-wave intensity, $\alpha_0 = m\eta$ is the small-signal absorption coefficient, m is the molar concentration, η is the extinction coefficient, and L is the thickness of the medium. The saturated-wave intensity based on the experimental conditions was $I_{sat} \fallingdotseq 250$ MW/cm^2. The reflectance R is likewise described by the theoretical curve of Equation (6.10). The maximum reflectance obtained was approximately 50%.

The existence of an optimum degree of concentration as seen in Figure 6.18 is observed not only for DODCI, but for CS_2 in hexane,[11] organic azo dyes in methanol, and ethanol (benzo-purpurin 4B and chrysoidin).[17] Eosin and erythrosin were used when the phase-conjugate wave reflectance was measured by a theoretical model expanded into a three-level atomic system; good agreement between experiment and theory was seen in these examples.[13]

c. Liquid Crystals

We will use MBBA (*p*-methoxybenzylidene *p'*-*n*-butylaniline) as an example of a liquid crystal.[21] MBBA is a nematic liquid crystal at room temperature. It was sealed in a quartz cell 5.7 cm long that contained an N_2 atmosphere, and was thermally regulated to within $\pm 0.2°$. The setup was a counterpropagating configuration. Vertically- and horizontally-oriented single-mode Q-switched ruby lasers (wavelength: 0.694 μm; pulse width: approximately 20 ns; pulse energy: 10 mJ) were used as light sources. The laser beam emerging from the source was concentrated at the center of the test cell by a mirror having a radius of curvature of 6 m. The distance from the light source to the test material was 12 m. As a result, there was a time delay of 80 ns corresponding to the round-trip distance of 24 m. Under these circumstances, the light source is prevented from becoming unstable due to beam reflection from the cell. The polarized probe wave (p wave) was orthogonal to the two pump waves (s waves), and was injected into the cell. The light beams were superposed; the angle of intersection between the probe and pump waves was held to approximately 10 mrad in order to obtain high interaction. The output phase-conjugate light was p-polarized.

The reflectance of MBBA can be accurately expressed by $R = \tan^2 (0.209 P_f)$ (P_f: pump-wave pulse energy, $P_f \leq 3$ mJ). The reflectance varies with temperature and reaches a maximum at 55°C. If the temperature drops, the reflectance decreases because of an increase in scattering loss as the phase-change point (10°C) is approached. Although the nonlinearity estimated from this data was lower than the expected value, the effect is that self-focusing occurs because of the high intensity, resulting in a decrease in beam superposition. Consequently, circular polarization was used to decrease the effects of self-focusing. If $P_f = 4.4$ mJ when a pump wave is circularly polarized but rotating in the opposite direction, the reflectance is 230%.

6.4 Special Mechanisms for Four-Wave Mixing Experiments

We explain in this section experiments in phase-conjugate light production by two-photon resonance absorption, forward four-wave mixing, and photon echoes, examples of which are shown in Table 6.5.

(1) Two-Photon Resonance Absorption

Since polarization used for the production of phase-conjugate light is mostly based on one-photon absorption, the upper- and lower-level carrier distributions are related to the production of phase-conjugate waves. We will show here an example of phase-conjugate light production based on two-photon resonance absorption related to interlevel coherence for CuCl.[47] The light source is an N_2-excited dye laser (pulse width: 5 ns; pulse energy: 20 μJ) in a counterpropagating configuration. The angle of intersection of the pump and probe waves inside the medium is

Table 6.5 Characteristics of special mechanisms in phase-conjugate light generating experiments

Substance name Symbol (unit)	Wave-length (μm)	Laser		Beam power	Nonlinearity $\chi^{(3)}$ (esu)	Characteristic	Bibliographic reference
		Type	Operating mode				
CuCl	0.389	Dye	Pulse	20 μJ	3×10^{-7}	Two-photon absorption	47
Polysilane	0.532	Nd : YAG (SHG)	Mode-locked	—	1.6×10^{-12}	Plasma effect	48
PTS polymer	0.652	Dye		100 W	9×10^{-9}	Macromolecular single crystal	49
α-(BEDT–TTF)$_2$I$_3$	0.65	Dye		80 μJ	5×10^{-8}	Organic metal	50
GaAs–MQW	0.852	Dye	Mode-locked		0.06	Exciton resonance	51, 52
GaInAsP	1.5	F nucleus	Mode-locked	3 nJ	3.8×10^{-3}	Resonance absorption	9

approximately 1°. The thickness of the test material is 1.25 μm, the measured temperature is 1.8 K and the pump intensity is $I_f = 1$ MW/cm^2. Figure 6.19 is a plot of reflectance versus wavelength for CuCl. In the diagram, the wide absorption peaks in the neighborhood of 387.5 nm are based on one-photon absorption; they start at the exciton-absorption edge (387 nm). On the other hand, the sharp peaks in the vicinity of 389 nm are due to biexciton two-photon resonance absorption; they appear only for the *xxxx* and *xyyx* configurations and have almost equal reflectances. Based on theoretical examinations, biexcitons are symmetric and, since they satisfy $\chi^{(3)}_{xxxx} = \chi^{(3)}_{xyyx}$, the latter absorption peak is thought to be based on two-photon coherence. From reflectance-versus-wave-intensity data, the optical nonlinearity originating from two-photon absorption was found to be $\chi^{(3)}_{xxxx} = 3 \times 10^{-7}$ esu (4.2×10^{-15} m^2/V^2). Similar data were also obtained from another experiment.[53]

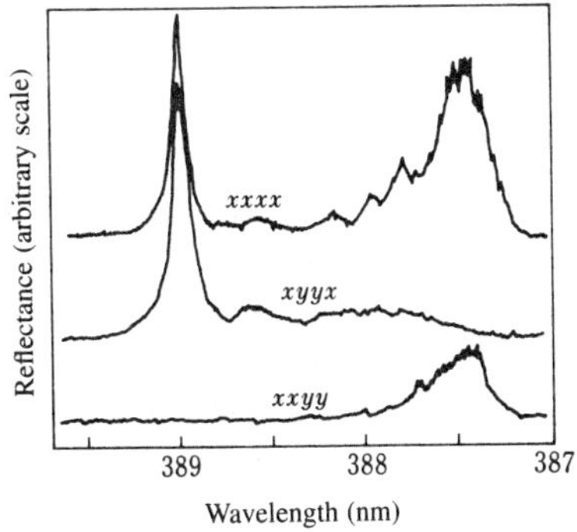

Figure 6.19 Reflectance versus wavelength for CuCl, based on two-photon resonance absorption. *(After Chase, Claude, Hulin and Mysyrowicz)*[47]

Thickness of test material: 1.25 μm; pump-wave intensity: 1 MW/cm^2; temperature: 1.8 K.
The *x* direction is perpendicular to the scattering plane; the *y* direction lies in the scattering plane.

Two-photon resonance absorption has also been observed with SF$_6$ (wave-length: 10.6 μm)[54] and with Rb vapor (wavelength: 760 nm).[55]

(2) Forward Four-Wave Mixing

We now take up the multiple quantum well (MQW) structure as an example that uses excitonic nonlinearity.[51,52] The test material was a periodic deposition of 9.6 nm-thick GaAs and 9.8-nm-thick $Ga_{0.72}Al_{0.28}As$ on a GaAs substrate (65 coatings). The thickness of the MQW was 1.26 μm. Following the removal of the substrate by selective etching, approximately 1 mm^2 of test material was bonded to sapphire to avoid thermal effects. Measurements could be made using a degenerate forward four-wave mixing configuration. The light source was a dye laser (wavelength: 770-870 nm; pulse width: 6 ps) and the probe wave had a time delay τ relative to the pump wave. The angle formed by the pump and probe waves was 16° outside the medium. Separate choppers were used for the pump and probe waves, and coincidence detection was used to increase the sensitivity. The self-diffracted phase-conjugate wave was measured by an optical detector.

When the phase-conjugate wave intensity for GaAs / GaAlAs MQW versus wave energy was measured, a peak was found at 1.457 eV (wavelength: 852 nm).[51] The pump- and probe-wave powers were 300 μW and 60 μW, respectively. When the relative delays τ for both waves were added, a phase-conjugate wave was observed only for a pulse width corresponding to $\tau = 6$ ps. By measuring the diffracted-wave intensity and the probe- and pump-wave transmittivity simultaneously, the nonlinearity originating in this case from excitonic resonance due to heavy holes could be clarified. Estimating the magnitude of the nonlinearity from this data, we obtain an unusually large value of $\chi^{(3)} \approx 0.06$ esu (8.4×10^{-10} m^2/V^2). We obtain an absorption coefficient of $\alpha = 1.2 \times 10^4$ cm^{-1} and a carrier lifetime of approximately 20 ns from the same test material.

There are cases of PTS macromolecular single crystals as examples that make use of nonlinearity due to conjugate π electrons.[49] The thickness of the test material was approximately 1 mm and the surface area was 50 mm^2. The light source was a cavity-dumped dye laser (wavelength: 652-702 nm; pulse width: 6 ps; optical power: 100W) which was concentrated to a beam spot of approximately 20 μm on the test material. The polarization of the waves was held parallel to the major axis of the PTS single crystal. The intensity I_c of the generated phase-conjugate wave was proportional to the product of the pump waves and the probe wave, $I_f I_b I_p$. A resonance peak appeared in the neighborhood of 650 nm when reflectance versus wavelength was measured. Because of absorption, the reflectance — that is to say, $\chi^{(3)}$ — as well as the absorption decreased as the wavelength shifted away from the resonance domain and became longer. In the non-resonant neighborhood of 700 nm, $\chi^{(3)} \approx 5 \times 10^{-10}$ esu. As shown in

Table 7.5, if the polarization of the light wave is perpendicular to the major axis of the PTS single crystal, the value of $\chi^{(3)}$ dramatically decreases and descends below the limits of detection in this experiment.

(3) Photon Echoes

An experiment using Na vapor was performed using the setup shown in Figure 6.20.[56] The light source was an Ar ion exciton dye laser (wavelength: 589 nm; power: 70 mW) that was passed through a dye-laser (rhodamine 6G) amplifier. The pulse width was decreased from 10 ns to 5 ns as a result of reflection by a diffraction grating (2380 lines/mm). These pulses were divided into two pulses having a 1:4 ratio — a first pulse ($\pi/2$ pulse) and a second pulse (π pulse). The delay between the pulses was set at 8 ns. The curvatures of lenses L_1 and L_3 (L_2 and L_3) were adjusted, the first (second) pulse's wave front was adjusted to satisfy Equation (4.54), and the waves were then injected into the Na cell (heated to 170°C). A photon echo was produced 8 ns after the second pulse — which had a wave intensity of 5 W/cm^2 — was injected. The photon echo had an intensity of 50 mW/cm^2 and was phase-conjugate to the first pulse.

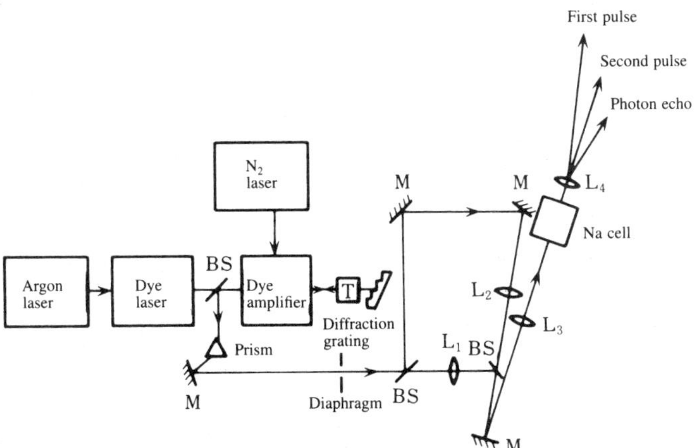

Figure 6.20 Experimental setup using photon echoes for generating phase-conjugate light. *(After Griffen and Herr)*[56]

BS: beam splitter; M: mirror; T: telescope; L: lens.

Although two pulses were injected in the example described above, experiments producing photon echoes have been performed with three injected pulses and where Yb vapor (wavelength: 555 nm)[57] and Pr^{3+}:LaF$_3$ (wavelength: 478 nm)[58] were used.

pulses and where Yb vapor (wavelength: 555 nm)[57] and Pr^{3+}:LaF_3 (wavelength: 478 nm)[58] were used.

6.5 Transient Response Characteristics of Four-Wave Mixing

The discussion in this section will focus on the interesting experimental results obtained when the optical pulse width is shorter than the response time of the nonlinear medium. These types of experiments are used mainly in solid-state physics research. Table 6.6 assembles examples that use short pulses for measur-

Table 6.6 Uses of transient response characteristics in the production of phase-conjugate light

Substance name Symbol (unit)	Wavelength μm	Laser		Pulse width	Target transient phenomenon	Time	Bibliographical reference
		Type	Beam power				
PTS polymer	0.652	Dye	2.4 kW	300 fs	Excited-state lifetime in resonance domain	1.8 ps	59
	0.723				Excited-state lifetime in nonresonance domain	<300 fs	
GaAs	0.813	Parametric ($LiIO_3$) Nd : YAG (SHG) excitation	10 μJ	7 ps	Excited-state lifetime in band-filling effect 77K	8.2 ps	60
Ge	1.06	Nd : YAG	10 mJ/cm^2	30 ps	Excited-state lifetime in band-filling effect	250 ps	61
BaTiO$_3$	0.532	Nd : YAG (SHG)	1–15 mJ/cm^2	30 ps	Free-carrier recombination time	<100 ps	62
BSO	0.532	Nd : YAG (SHG)	1 mJ	3 ns	Recombination time from the electron conduction band	50 ns	63

ing phase-conjugate-wave transient response, and explains the transient phenomena of various media. We next detail several examples of forward and back four-wave mixing.

(1) Transient Response Characteristics of Back Degenerate Four-Wave Mixing

We will consider as an example an experiment that was performed using a counterpropagating configuration.[64] The light source was a Q-switched Nd:YAG laser (wavelength: 1.06 μm; pulse width: 10 ns), with the lengths of the wave paths from the light source to the test material adjusted to be equal. Figure 6.21(a) shows the phase-conjugate pulse waveform for CS_2. CS_2 can readily be used as the standard test material for measuring the absolute value of $\chi^{(3)}$. The relaxation time for CS_2 is less than 1 ps, and since this is narrower than the width of the wave pulse, an almost symmetric waveform is obtained, as explained in Section 4.4. The ringing on the right-hand side is caused by the measuring circuit. Let us assume from Equation (4.59) that the wave pulse has a Gaussian waveform. Assume also that we are dealing only with fast responses and that the relaxation time $\tau_{gR} \to 0$. This results in good agreement of the resulting theoretical values

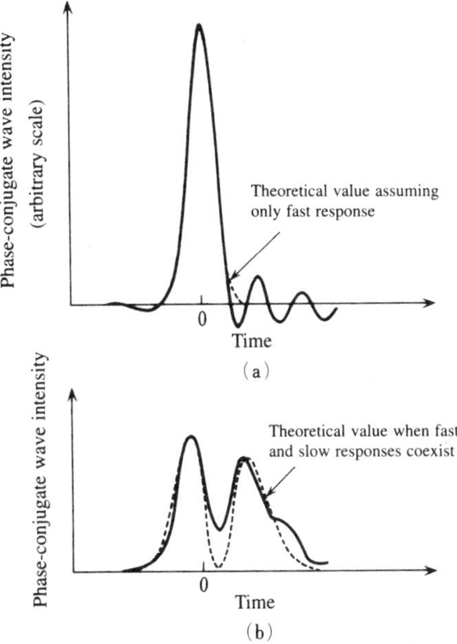

Figure 6.21 Phase-conjugate temporal wave forms for pulse-wave injection in degenerate four-wave mixing. *(After Smith, Tomlinson, Eilenberger and Maloney)*[64]

(a) CS_2 Solid curve: experimental values; dashed curve: calculated values as $\tau_{gR} \to 0$.
(b) Retinal (when fast and slow responses coexist) Solid curve: experimental values;
Dashed curve: calculated values when the ratio of the slow to the fast response component is $F = 0.8$.

configuration as in (a). Under these circumstances, two characteristic peaks appear because of retardation of the response speed due to thermal effects. The dashed curve represents the values calculated from Equation (4.59) as the fast relaxation time $\tau_{gR} \to 0$ and as the slow response time $\tau_{gT} \to \infty$; the ratio F of the slow to the fast response component is 0.8. The calculated and experimental values agree quite well. Measurements for retinal and β-carotene were also made and the values for $\chi^{(3)}$ of the same order as for CS_2 were obtained.

Although the next example also uses a counterpropagating configuration, this time a delay circuit was inserted in the path of the pump waves in order to independently vary the path lengths of the forward pump E_f, the back pump E_b and the probe E_p from the light source to the test material.[65] In addition, polarized photons and a half-wave plate were inserted in each optical path in order to selectively generate a diffraction grating contributing to the production of the phase-conjugate wave. The light source was a mode-locked Nd:YAG laser (wavelength: 1.06 μm; pulse width: 33 ps; energy: 5 mJ). The test material was a polydiacetylene (structural formula: $R–(C–C\equiv C–C)_n–R'$), the so-called poly-m-BCMU ($R=R'=–(CH_2)_m–OC–ONHCH_2COO(CH_2)_3CH_3$; $m = 3,4$) dissolved in an organic medium. Test material of 1 mm thickness was placed in a quartz cell. The angle between E_p and E_f was set to 6° at the test material's surface. After the

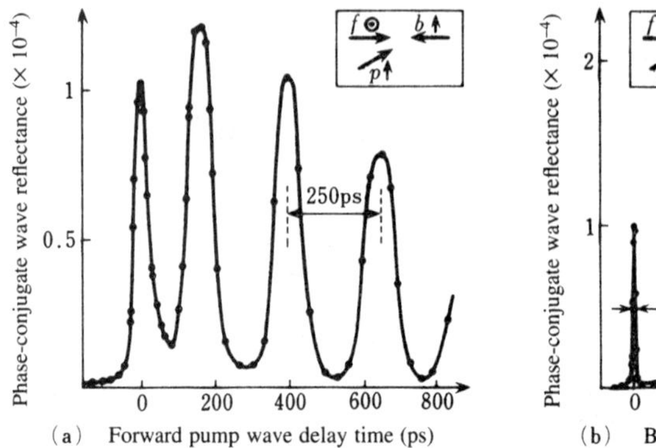

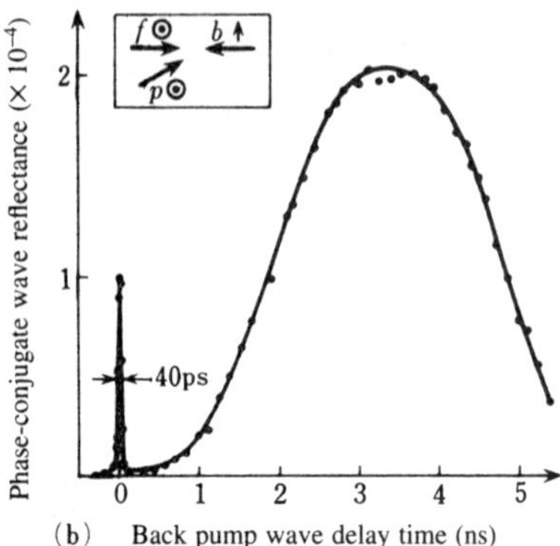

Phase-conjugate wave reflectance (× 10⁻⁴)

(a)   Forward pump wave delay time (ps)

(b)   Back pump wave delay time (ns)

Figure 6.22 Phase-conjugate wave reflectance polarization versus probe wave delay time. *(After Nunzi and Grec)*[65]

(a) polarizations of the probe and forward pump waves are parallel and the forward pump wave is perpendicular to the other two waves
(b) polarizations of the probe and forward pump waves are parallel and the back pump wave is perpendicular to the other two waves
The test material is a solution of *m*-BCMU dissolved in DMF. Wavelength is 1.06 μm.

phase-conjugate waves were detected by an electron multiplier tube, they were averaged over ten data points by a boxcar integrator.

The polarizations of E_p and E_b are parallel in Figure 6.22(a) — a case of phase-conjugate wave reflectance when those two waves are orthogonally polarized with respect to E_f. E_f is delayed relative to E_p and E_b. The effects of acoustic oscillation caused by the narrow diffraction grating (spacing $\Lambda = 0.38$ μm) appear following the response due to fast electron transitions in the neighborhood of $\tau = 0$. The period of this acoustic oscillation is approximately 250 ps, about twice the interval up to the first peak. The major cause of acoustic oscillation is the medium's electrostriction; $T = 271$ ps if the time interval is calculated from $T = \Lambda/v_a$ (v_a: acoustic velocity; Λ: grating spacing) when $v_a = 1.4 \times 10^5$ cm/s. The calculated and experimental values agree well. On the other hand, the polarizations of E_p and E_f are parallel in Figure 6.22(b), a case of phase-conjugate wave reflectance when those two waves are polarized orthogonally to E_b. Here E_b is delayed relative to E_p and E_f. In this case the first peak of the acoustic oscillation caused by the coarse diffraction grating (spacing $\Lambda = 10$ μm) emerges approximately 3.5 ns after the response due to fast electron transitions in the neighborhood of $\tau = 0$. The acoustic oscillation in this case can likewise be explained by the electrostrictive effect of the medium.

Although a photorefractive medium has high equivalent nonlinearity, its slow response time has been considered to be a disadvantage. However, light sources with pulse widths in the sub-nanosecond range have been used recently and while the measurements are the same as those in Figure 6.22, it has become clear that the carrier recombination time is under 100 ps for $BaTiO_3$[62] and about 50 ns for BSO.[63] Response times in the dozens of picoseconds have been achieved for GaAs.[66] These values vary substantially from response times actually obtained (milliseconds to seconds for oxides; microseconds for semiconductors). We await future developments to fill this gap.

(2) Transient Response Characteristics of Two-Photon Absorption

We next show an example of two-photon resonance absorption.[67] The light source is an N_2-excited dye laser (wavelength: 579 nm; pulse width: 5 ns; power: 15 kW). The configuration was retroreflective with orthogonally polarized pump and probe waves. The wave could be easily separated since the polarizations of the phase-conjugate and back pump waves were perpendicular. The nonlinear medium was an atmosphere of Na mixed with Ne, with Na pressure maintained at 0.3 Torr. The wavelength was synchronized to the Na 3S $\rightarrow$ 4D two-photon transition. Figure 6.23 shows measured results for the phase-conjugate wave intensity I_c when the probe wave lagged the pump wave.

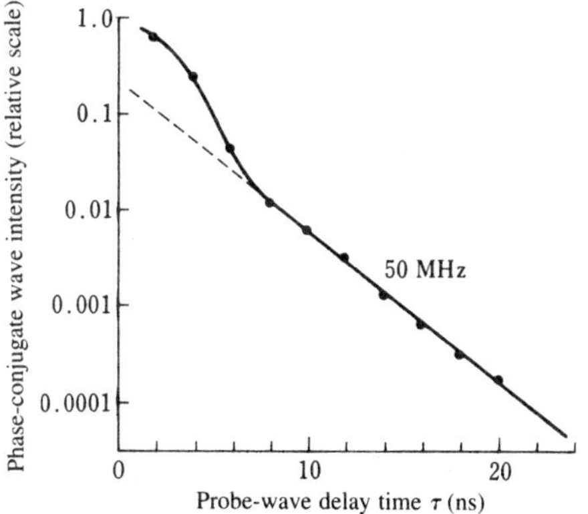

Figure 6.23 Transient response characteristics of two-photon absorption for Na vapor. *(After Liao, Economou and Freeman)*[67]

The medium is a gaseous mixture of Na (0.3 Torr) and Ne (0.8 Torr). Relaxation is due to collision processes.

The relaxation time can be obtained from the attenuation of I_c as shown by Equation (4.58). In addition, the full width at half maximum could be obtained from $\delta\nu = 1/\pi T_2$ (T_2: relaxation time). As a result, the relaxation due to Na-Ne collisions was found to correspond to a frequency of 50 MHz. The small hump in the neighborhood of $\tau = 0$ is the effect of the pump and probe waves being superposed.

(3) Transient Response Characteristics of Forward Degenerate Four-Wave Mixing

The experimental configuration is shown in Figure 6.24.[61] The light source was a mode-locked Nd:YAG laser (wavelength: 1.06 μm; pulse width: 30 ps; peak energy density on the test material surface: 10 mJ/cm²) in horizontal single mode.

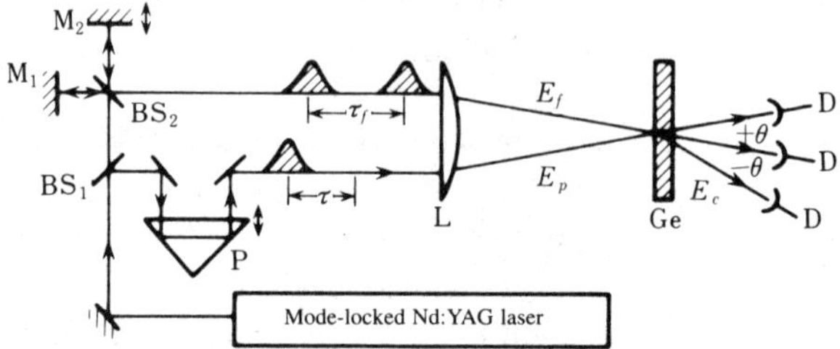

Figure 6.24 Setup for measuring transient response characteristics for forward degenerate four-wave mixing. *(After Smirl, Boggess and Hopf)*[61]

E_f: pump wave; E_p: probe wave; E_c: phase-conjugate wave; τ_f: delay time between the two pump waves; τ: probe wave delay time relative to the mid-point of the interval between the pump waves.

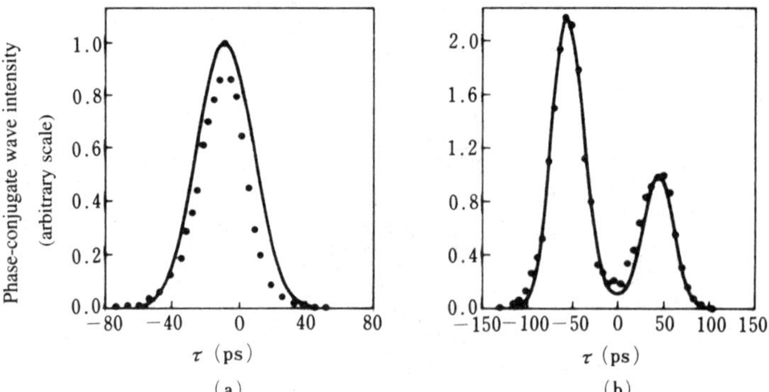

Figure 6.25 Relationship between phase-conjugate wave intensity in forward degenerate four-wave mixing and probe-wave delay time τ. *(After Smirl, Boggess and Hopf)*[61]

Test material: Ge; incident wave pulse width: 30 ps (FWHM); wavelength: 1.06 μm.
(a) Pump wave delay time $\tau_f = 0$ ps   (b) $\tau_f = 105$ ps
Solid curve represents theoretical values from Equation (4.59) with $\tau_g = 250$ ps.

The pulse width was measured by the SHG correlation method. The pulses from the light source were separated into two paths by beam splitter BS_1 and became pump wave E_f and probe wave E_p. The pump wave was subsequently broken down into two pulses by beam splitter BS_2, and a relative delay of τ_f was introduced between the two pump waves by movable mirror M_2. On the other hand, the probe wave was relatively delayed by time τ via a movable prism P at the center of the two pump waves. These light waves were concentrated on the test material to a 500-μm diameter spot by lens L. The angle θ formed by the pump and probe waves was 3.5° and the test material was Ge having a thickness of 6 μm. It is believed that the Ge functioned as a diffraction grating since the band-filling effect was the principal cause of absorption at a wavelength of 1.06 μm. Phase-conjugate wave E_c was generated here by self-diffraction of the pump wave and had the same frequency as the probe wave. The wave emerged at an angle of $-\theta$ to the pump wave.

Figure 6.25 shows probe-wave delay time τ versus phase-conjugate wave intensity. Figure 6.25(a) shows the measured values when the delay relative to the pump wave is $\tau_f = 0$; (b) shows the values for $\tau_f = 105$ ps.[61] The phase-conjugate wave intensity is nonsymmetric with respect to the delay time τ. As previously mentioned in Section 4.4, the reason for this is as follows. If the lifetime of the diffraction grating is much greater than the incident wave's pulse width, the pump wave self-diffracts as long as the diffraction grating continues to exist in any part of the pump wave that follows the superposed part of the pump and probe wave. Consequently, in the case of Figure 6.25(a), the intensity of the phase-conjugate wave increases less when $\tau < 0$ (probe leading pump) than when $\tau > 0$. Further, in the case of Figure 6.25(b), when the probe wave coincides with the preceding pump wave ($\tau \sim -50$ ps), the intensity of the phase-conjugate wave is higher for $\tau = -50$ ps than it is for $\tau = 50$ ps because the diffraction grating self-diffracts not only for the preceding pump wave but for the following pump wave. Incidentally, if the lifetime of the diffraction grating is much shorter than the pulse width, a symmetric effect related to the delay time τ is obtained since the diffraction grating exists only when the pump and probe waves are superposed. The solid line in Figure 6.25 resembles the nonlinear medium response function $W(t;\tau_g)$ of Equation (4.60), where the theoretical value is obtained for a degeneracy time of the diffraction grating $\tau_g = 250$ ps. In particular, the experimental and theoretical values agree well when τ_f is greater than the width of the pulse.

6.6 Three-Wave Mixing Experiments

A schematic diagram of an experiment for producing phase-conjugate light by three-wave mixing is shown in Figure 6.26.[68] The light source employed was a Q-switched Nd:YAG laser (used as the probe wave; wavelength: 1.06 μm; peak power: 30 – 100 kW), with a second harmonic generated by $LiIO_3$ (used as the pump wave; wavelength: 0.53 μm; peak power: ~1 kW). The powers of both waves were set to an appropriate level, and a cell was inserted to adjust the

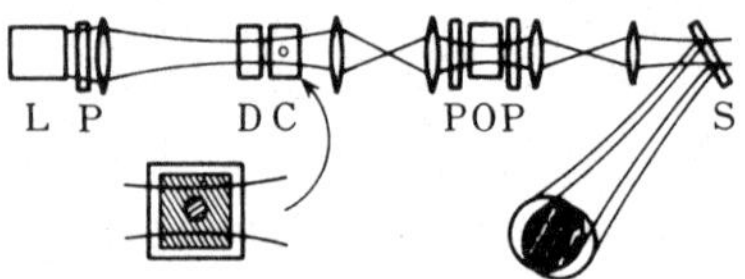

Figure 6.26 Experimental setup for phase-conjugate light generation by three-wave mixing. *(After Avizonis, Hopf, Bomberger, Jacobs, Tomita and Womack)*[68]

L: Q-switched Nd:YAG laser; P: polarizer; D: LiIO$_3$ doubler; C: refraction adjustment cell; O: optical parametric amplifier; S: wedge shearing interferometer.

refractive index after the polarization orientations of both waves were adjusted to be the same. Accordingly, there was zero phase shift at 0.53 μm and a 3/4-wavelength phase shift at 1.06 μm. This happened because the idler wave became precisely phase-conjugate to the probe wave when there was zero pump-wave phase shift. An optical parametric amplifier (OPA) was fashioned from LiCHO$_2$:H$_2$O and placed between the orthogonal polarizers. Only the idler wave emerged from the OPA under Type II phase-matching conditions, that is, the pump and signal waves had the same polarization orientation, and the idler wave (1.06 μm) was orthogonal to them.

When these waves were injected into a wedge shearing interferometer, interference fringes were observed with an infrared vidicon. Straight fringes were obtained when there was no phase shift; an S-shaped pattern was obtained when there was distortion. That the idler wave became phase-conjugate to the signal wave could be confirmed by its change in direction, which was exactly opposite to the change in direction of the fringes.

It is important to the experiment that the idler wave intensity is higher than the background wave intensity level coming from the crossed Nicols. This is because LiCHO$_2$:H$_2$O is a biaxial crystal and subtle adjustment is required for high extinction by the crossed Nicols. The phase-conjugate wave intensity was approximately 0.5% that of the signal wave.

6.7 Stimulated Scattering Experiments

As previously mentioned, phase-conjugate light was discovered in experiments with stimulated Brillouin scattering (SBS). In the early stages, compounds such as methane gas, CS$_2$, and SF$_6$ were tried out in the experiments. However, both SBS and stimulated Raman scattering (SRS) were accompanied by frequency shifts and there were subsequently few experiments conducted due to faulty

Table 6.7 Characteristics of experiments in phase-conjugate light production with stimulated wave scattering

Substance name Symbol (unit)		Wavelength λ (μm)	Laser		Beam power	Stimulated scattering gain function g ($\times 10^{-5}$m/MW)	Reflectance %	Bibliographical reference
			Type	Operating mode				
Hexane, alcohol	SBS	0.351	XeF	Injection synchronous	50 mJ	—	70	69
Methane	SBS	0.694	Ruby	Pulse	1.3 MW	90	25	70
CS_2	SBS	0.694	Ruby	Q-switched	25 mJ	128	10–70	71, 72
CCl_4	SBS	1.06	Nd : YAG	Pulse	20 J	6	—	73
Xe	SBS	2.91	HF	Pulse	3 J	44	50	74
CS_2	SRS	0.694	Ruby	Pulse	29 MW	24	—	75, 76

SBS: stimulated Brillouin scattering; SRS: stimulated Raman scattering

phase-conjugate characteristics. Several examples of the experiments are shown in Table 6.7. We next explain representative experimental results.

(1) Stimulated Brillouin Scattering

The configuration for a phase-conjugate light generating experiment using stimulated Brillouin scattering is shown in Figure 6.27, along with phase-compensation data.[71] The light source was a vertical single-mode Q-switched ruby laser (wavelength: 0.694 μm; pulse width: 17 ns). The beam was optically amplified to an energy of 25 mJ immediately in front of the Brillouin cell. The nonlinear medium was CS_2 contained in a 1-m-long, glass-tube, multimode waveguide having an inside diameter of 2.5 mm, producing a large interaction length. Since the refractivity of CS_2 is higher than that of the glass, there was complete reflection within the waveguide. The back-scattered waves generated by the Brillouin cell were slightly shifted in frequency and followed the original path. They were recorded by a camera after reflection by a beam splitter. Since a phase-conjugate component is present in stimulated Brillouin scattering, even if an aberrating substance is inserted along the wave path, it is clear that beam-spreading compensation occurs, as shown in Figure 6.27(b). The beam was spread to 6.6 mrad by inserting an aberrating substance and was restored to 0.44 mrad by compensation of the phase-conjugate wave. These values are about the same as those for the original beam width and also agree with the calculated values for the diffraction limit. This means that the beam width is diffraction-limited if the phase-conjugate mirror is used as a reflective mirror, even when there is an aberrating substance in the wave path. In reality, this approach is applied to optical resonators.

Experiments with nuclear fusion and lithography using ultraviolet light have also been performed.[69]

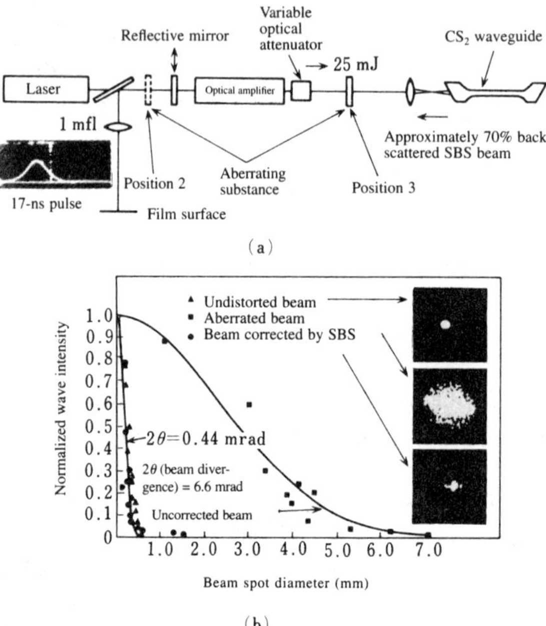

Figure 6.27 Use of stimulated Brillouin scattering to compensate phase distortion due to aberrating substance. *(After Wang and Giuliano)*[71]

(a) Experimental setup for generating phase-conjugate light; wavelength: 0.69 μm
(b) Phase correction data: beam-spreading angle

(2) Stimulated Raman Scattering

The Stokes frequency shift produced by stimulated Raman scattering is $100 - 1000$ cm^{-1}, three to four orders of magnitude greater than that in SBS. CS_2 and the same ruby laser light source were used; SRS exhibited more phase-conjugate wave quality degradation than SBS in an experimental comparison. In addition, there was high degradation when the beam's cross-sectional area was large.[75]

(3) Brillouin-Enhanced Four-Wave Mixing

Assume that a probe signal wave is produced by stimulated Brillouin scattering. Assume also two counterpropagating pump waves having the same frequency as the probe. The compound wave that results from injection into the same type of SBS-producing nonlinear medium is called Brillouin-enhanced four-wave mixing.[77] The interference bands thus formed by the probe and back pump waves migrate at acoustic-wave velocity. Although the forward pump wave is diffracted by the interference fringes to produce a phase-conjugate wave, with SBS the back pump wave yields a component with the same frequency as the probe wave, accompanied by an improvement in reflectance over normal degenerate four-wave mixing. Assuming that acetone is the nonlinear medium and that an Nd glass laser is the light source (wavelength: 1.06 μm; pulse width: 33 ns), and also assuming that the angle formed by the pump and probe waves is 1.5×10^{-2} rad, an extremely high reflectance of 7×10^5 can be obtained.[78] CS_2 and $TiCl_4$ have also been used as nonlinear media.

7.

Phase-Conjugate Light Generating Media

Nonlinear media are used for producing phase-conjugate light. The characteristics of nonlinear media have been studied through self-focusing effects, the generation of third harmonics, and stimulated scattering. In this chapter, we will first explain the characteristics required of media when the objective is the production of phase-conjugate light. We will then explain in great detail the nonlinear polarization that was discussed in terms of parameters in Section 3.1. Guidelines for media will be covered in Section 7.1 and, after the origins of nonlinear polarization are covered in Section 7.2, Sections 7.3 – 7.8 will give expressions for the nonlinearity for each medium in particular.

7.1 Guidelines for Phase-Conjugate Light Generating Media

Nonlinear media make use of wavelength conversions, optical bistability, and optical switching elements, among other phenomena, where the required conditions depend on the objective. Media guidelines for the production of phase-conjugate light are enumerated below.

(i) A high third-order nonlinearity $\chi^{(3)}$ should be selected to increase reflectance, as shown by Equation (3.28).

(ii) In order to effectively extract phase conjugate light outside the medium, it is necessary that the medium be transparent to some degree. However, the region of transparency becomes a problem. For example, $\chi^{(3)}$ is large near the absorption edges in semiconductors because free electrons and excitons contribute to increasing $\chi^{(3)}$. The visible spectrum cannot be used since multiple absorptive limits usually lie within the infrared region. Since the absorption coefficient α is large in CuCl and multiple-quantum well-structured GaAs, the thickness of the medium must be in the μm range.

(iii) Assuming a continuous pump wave, as explained in Section 4.4, if the medium's response time τ is shortened when the probe wave is pulsed, the phase-conjugate wave becomes faithfully phase-conjugate in relation to the probe wave.

(iv) In order for phase-conjugate light to be generated in a saturable medium by the smallest wave intensity possible, the medium with the smallest saturation intensity I_{sat} should be selected.

In general, these parameters cannot be independently determined. $\chi^{(3)}/\alpha\tau$ assumes a constant value independent of the medium (refer to Table 7.4). That is, if the response time τ is short, the tendency is for the nonlinearity $\chi^{(3)}$ to become small. A small τ is incompatible with a large $\chi^{(3)}$. Consequently, when selecting a medium it is necessary to make certain that these characteristics satisfy the objectives.

7.2 Sources of Nonlinear Polarization

Sources of nonlinear polarization are broadly classified as resonant and nonresonant. When a medium is sufficiently far from the resonance absorption line, the principal causes of nonlinear polarization are — among other things — (i) the optical Kerr effect, (ii) molecular redistribution, (iii) nonlinear electron polarization, (iv) electrostrictive effects, (v) thermal effects, (vi) photorefractive effects, and (vii) radiation pressure. At resonance, on the other hand, (viii) saturation effects can be observed. Which mechanism is most effective depends, of course, chiefly on the type of substance, as well as the wavelength and pulse width of the incident waves, the temperature, and other conditions. Principal factors and representative substances are shown in Table 7.1, along with typical nonlinearity and response times. We explain the essence of each effect in what follows.

The refractive index n of a medium can be represented in SI units by the Lorenz-Lorentz relationship[1]

$$\frac{(n^2-1)}{(n^2+2)}=\frac{1}{3}\rho_0\langle\alpha_p\rangle \tag{7.1}$$

Table 7.1 Chief mechanisms of nonlinear polarization

Mechanism	Principal medium	$\chi^{(3)}$ (esu)	Response time (sec)
Optical Kerr effect	Na vapor	$10^{-13} \sim 10^{-12}$	$10^{-12} \sim 10^{-11}$
Molecular reorientation	CS_2, nitrobenzene	$10^{-14} \sim 10^{-12}$	10^{-13}
Nonlinear electron polarization	CCl_4, Ar	$10^{-15} \sim 10^{-14}$	10^{-15}
Constrained electrons in anharmonic potential	Ge, Si, GaAs, CdS, CdSe	$10^{-12} \sim 10^{-10}$	$<10^{-15}$
Free-electron effect due to non-parabolicity of the conduction band	n-InSb, n-InAs, CdHgTe	$10^{-10} \sim 10^{-7}$	$<10^{-13}$
Band filling	InSb, HgCdTe	$10^{-5} \sim 10^{-2}$	10^{-8}
Multiquantum-well effect in crystals	CdS_xSe_{1-x} doped glass	10^{-8}	10^{-9}
Photorefractive effect	Dielectrics ($BaTiO_3$, BSO, SBN, $LiNbO_3$, $LiTaO_3$)	$(10^{-5} \sim 10^{-4})$	$10^{-3} \sim 1$
	Semiconductors (GaAs, InP, CdTe)		10^{-6}
Conjugate π electrons	Macromolecular single crystals (PTS polymers, TCDU polymers)	$10^{-10} \sim 10^{-9}$	$<10^{-12}$
Thermal effects	CdHgTe, organic compounds as solvents	$10^{-6} \sim 10^{-5}$	$10^{-1} \sim 1$
Electrostrictive effects	Lucite, CS_2, nitrobenzene	$10^{-13} \sim 10^{-12}$	$10^{-9} \sim 10^{-8}$
Radiation pressure	Spherical particles in a liquid (polystyrene spheres, latex spheres)	$10^{-8} \sim 10^{-7}$	10^{-1}

where ρ_0 [m^{-3}] is the average molecular density and $<\alpha_p>$ [m^3] is the average molecular polarization. Accompanying a change in $<\alpha_p>$ is a change in the refractive index described in (i), (ii), and (iii) above, which appears in gases and liquids. A change in the refractive index accompanies a change in ρ_0 for (iv) and (v). The optical Kerr effect reorients the molecules (of CS_2, for example) in accordance with the anisotropic polarization resulting from the incident electric field, thus changing the refractive index. The redistribution of molecules is an interaction among the dipole moments induced by spherical molecules and effects a change in the molecular distribution. Nonlinear electron polarization is produced in a nonsymmetric medium and is caused by electron-cloud distortion due to the electric field. The electrostrictive effect changes molecular density due to changes in pressure; this effect is dramatic in solids and liquids. The thermal effect localizes increases in temperature because wave energy absorbed by the medium is dissipated non-radiatively, thus producing a change in the refractive index via a change in density. A longer response time is a typical effect of the change in density. In particular, thermal effects slow response time considerably even though the refractive-index change is large.

The photorefractive effect can be seen in dielectric substances and semi-conductors when, due to the wave energy, carriers migrate from the donor level to the conduction band, creating a distribution of electric space charge. The charge distribution sets up an electric field and produces a change in the refractive index due to the Pockels effect. Radiation pressure is produced when globular particles (such as polystyrene and latex) are suspended in a liquid. Although the globular particles themselves do not possess nonlinearity, they can have stable localized potential due to the radiation pressure of the light beam, which results in nonlinearity.

Absorption effects — for example, when incident waves correspond to certain frequencies between two atomic or molecular levels — excessively increase the number of excited carriers, the result of which is saturation during absorption, accompanied by a change in the refractive index.

There is a detailed literature[1-3] that discusses the long-known mechanisms of nonlinear polarization. In addition, there is a rich bibliography[2,4] addressing numerical examples of nonlinearity. Although the nonlinearity values shown in Table 7.1 vary from one system of units to another, conversion factors are shown in Appendix A.

7.3 Gases

Atomic vapors are easily used for the production of phase-conjugate light. If a wave having a frequency ω near the resonant frequency of the medium (ω_R) is injected, the lower-level atoms are excited by the absorbed wave energy and a redistribution of atoms occurs between energy levels. There is a change in the refractive index due to the changing effects of dispersion that accompany the absorption. This phenomenon is called a self-induced index change. Since saturation effects are produced by this phenomenon, the medium is generically called saturation-absorptive. In addition to atomic vapors, organic dyes and electron transitions in semiconductors fit the saturation-absorption model well.

A change in refractive index can be investigated through classical dispersion theory by assuming a two-level atom such that the excited system returns to thermal equilibrium with a time constant τ.[4,5] The nonlinear refractive index n_2 and the intensity-dependent absorption coefficient α in SI units are given by the following equations.

$$n_2 = \frac{n_0 c^2 \varepsilon_0}{2\omega} \cdot \frac{\alpha_0}{I_{sat}^0} \cdot \frac{(\omega - \omega_R)\, T_2}{[1 + (\omega - \omega_R)^2 T_2^2]^2} \quad [\text{m}^2/\text{V}^2] \tag{7.2a}$$

$$\alpha = \alpha_0 / \{[1 + (\omega - \omega_R)^2 T_2^2][1 + (I/I_{sat})]\} \quad [\text{m}^{-1}] \tag{7.2b}$$

$$\alpha_0 = (\omega/n_0 c)\,[\delta N \mu_{\parallel}^2 T_2/(\hbar \varepsilon_0)] \tag{7.3a}$$

$$I_{sat}^0 = (n_0 c \varepsilon_0/2)\,[\hbar^2/(\mu_{\parallel}^2 \tau T_2)] \tag{7.3b}$$

$$I_{sat} = I_{sat}^0\,[1 + (\omega - \omega_R)^2 T_2^2] \tag{7.3c}$$

where α_0 is the small-signal absorption coefficient, I_{sat}^0 is the saturation intensity at the absorption center, I_{sat} is the saturation intensity with detuning, I is the incident wave intensity, T_2 is the phase relaxation time, $\mu_{\parallel}$ is the dipole moment parallel to the electric field, δN is the population difference between upper and lower levels at equilibrium, n_0 is the linear refractive index at a wavelength sufficiently far from resonance, and c is the speed of light in vacuo. The expressions above for α_0 and I^0_{sat} are equivalent to Equations (4.14) and (4.11) of semi-classical theory. The nonlinear refractive index depends on, among other things, detuning between the frequency of the wave and the resonant frequency of the medium, the saturation intensity, and the absorption coefficient.

In a two-level atomic system, τ is the lower-level relaxation time and, simultaneously, the upper-level lifetime. In an actual system, although excited atoms (molecules) suddenly relax to a neighboring level when multiple levels exist, there is incomplete relaxation to the original lower level because of thermal fluctuations, making the relaxation time normally slower than 10^{-6} seconds. However, the lifetime can be under 10^{-6} seconds if an appropriate buffer gas or solvent is used.

T_2 in Equation (7.3b) should be made as large as possible in order to decrease the saturation intensity. Normally, $T_2 < \tau$, and T_2 can be made as close as desired to τ. Here, an appropriate buffer gas should be selected in the case of a gas. Since the molecules are randomly distributed, $T_2 < 10^{-10}$ seconds. It has been verified from the results found for each type of atomic vapor, molecular gas, organic dye, metal ion, and semiconductor that it is possible to achieve a low saturation intensity in a low-pressure atomic vapor.[4] For example, $I^0_{sat} = 49$ W/cm^2 has been obtained at 589 nm in the case of Na vapor. Moreover, if we assume that $(\omega - \omega_R)T_2 = 10$ (0.5 cm^{-1} separation from the absorption line), then we obtain $I_{sat} \approx 5$ kW/cm^2 from Equation (7.2a), $n_2 = 3.1 \times 10^{-11}$ cm^2/W (7.5 $\times$ 10^{-9} esu) when an Na cell with length $L = 6$ cm is assumed to be at 150°C. The linear refractive index $n_0 \approx 1$.

Na vapor, Rb vapor, methane, SF$_6$, and NH$_3$ have been used in experiments in phase-conjugate light production. If a magnetic field is impressed in order to cancel the level degeneracy in Rb vapor, the nonlinear refractive index changes as a function of the direction of circularly polarized light and $n_2 = -5.4 \times 10^{-10}$ cm$_2$/W (-1.3×10^{-7} esu; wavelength: 795 nm) is obtained at resonance.[6]

7.4 Particles in Liquids and Solutions

(1) Liquids
a. Optical Kerr Effect

We will first explain molecular reorientation. Consider molecules having anisotropic polarization and assume that the polarization parallel (perpendicular) to the molecular axis of symmetry is $\alpha_{\parallel}$ ($\alpha_{\perp}$), where $\alpha_{\perp} > \alpha_{\parallel}$. If we further assume that θ is the angle made by the incident wave's electric field E and the molecular axis of symmetry, then the average polarization in the direction of E can be considered to be

$$\langle \alpha_p \rangle = \alpha_{\parallel} \langle \cos^2 \theta \rangle + \alpha_{\perp} \langle \sin^2 \theta \rangle = (\alpha_{\parallel} - \alpha_{\perp}) \langle \cos^2 \theta \rangle + \alpha_{\perp} \tag{7.4}$$

where $< >$ denotes the average over the statistical distribution. If Equation (7.4) is substituted in Equation (7.1) and we take the lowest-order term of the Taylor expansion, we obtain the following equation for the nonlinear refractive index:[1]

$$n_2 = \frac{16\pi^2 \rho_0 \varepsilon_0}{45 n_0} \left(\frac{n_0^2 + 2}{3} \right)^4 \frac{(\Delta\alpha)^2}{k_B T} \quad [\text{m}^2/\text{V}^2] \tag{7.5}$$

where n_0 is the linear refractive index, ρ_0 [m^{-3}] is the average number density of molecules, $\Delta\alpha = \alpha_{\parallel} - \alpha_{\perp}$ [m^3], ε_0 is the permittivity of free space, k_B is Boltzmann's constant, and T is the absolute temperature. If Equation (7.5) is computed for CS_2, $n_2 \approx 1.4 \times 10^{-20}$ m^2/V^2 (1.3×10^{-11} esu), which compares well with the measured value $n_2 = 1.3 \times 10^{-20}$ m^2/V^2 (1.2×10^{-11} esu) (wavelength: 694.3 nm).

CS_2 is a liquid that contains anisotropic molecules. In addition to having a high nonlinear refractive index, its absorption coefficient $\alpha \approx 10^{-3}$ cm^{-1} (wavelength: 632.8 nm) is extremely small. Its region of transparency extends from the visible to the infrared. In addition, although saturation occurs due to a completely uniform molecular orientation if the incident intensity is strong in a molecularly-oriented Kerr medium, saturation occurs for intensities only up to $I < 10^{12}$ W/cm^2 in CS_2.[4] Since CS_2 has this characteristic it can easily be used as a comparison standard (calibration standard) for assessing the magnitude of nonlinearity.

Nonlinear electron polarization is produced in media with symmetric molecules (e.g., CCl_4 and Ar) and is caused by electron-cloud distortion due to the electric field. It is characterized by a short relaxation time, which corresponds approximately to the Bohr orbital period (10^{-16} seconds). However, the nonlinearity is not so large, of the order of $\chi^{(3)} \approx 10^{-23}$ m^2/V^2 ($\sim 10^{-15}$ esu).

Incidentally, the refractive-index dispersion can be found from the Sellmeier equation

$$n^2 = a_1 + [a_2/(1 - \lambda_v^2/\lambda^2)] \qquad (\lambda : \text{nm}) \tag{7.6}$$

where α_1 and α_2 are the Sellmeier coefficients, and λ_v(nm) is the absorption wavelength. Sellmeier parameters for liquids often used in experiments as well as their nonlinear refractive indexes are shown in Table 7.2, along with refractive-index temperature coefficients.[2,4]

Table 7.2 Various parameters of liquids

Substance	a_1	$10a_2$	λ_v (nm)	$10^{11} n_2$ (esu) 694.3 nm	-10^5 (dn/dT)
CS_2	1.5630	9.512	212	1.20	80
CCl_4	2.040	0.562	342	0.02	50
Benzene	1.452	7.27	170	0.125	65
Nitrobenzene	1.5404	7.700	201	0.86	51
Acetone	2.1038	-2.904	211	0.043	53
Cyclohexane	1.7932	2.270	196	0.013	50
Toluene	1.9627	2.265	259	0.30	58

The Sellmeier parameters are given by Equation (7.6). n_2 is the nonlinear refractive index. dn/dT is the refractive-index temperature coefficient.

b. Thermal and Electrostrictive Effects

The following thermodynamic equation can be written for refractive-index fluctuations in terms of the medium temperature T and density ρ [kg/m³]:

$$\delta n = (\partial n/\partial T)_\rho \, \delta T + (\partial n/\partial \rho)_T \, \delta \rho \tag{7.7}$$

The first term on the right-hand side represents the thermal effect of entropic fluctuations; the second term represents the electrostrictive effect caused by fluctuations in density. We explain each of them in what follows.

A part of the light energy absorbed by the medium is transformed into heat in liquids containing dissolved organic compounds. These thermal fluctuations mirror the wave interference distribution. As a result, the grating thus created is called a thermal diffraction grating. The temperature distribution $T(t,r)$ at this time has been described by the dispersion equation[7,8]

$$\rho C_p \frac{\partial T}{\partial t} - K\nabla^2 T = \alpha_0 I_m \Phi \tag{7.8}$$

where C_p is the ratio of specific heats at constant pressure, K is the thermal conductivity, α_0 is the intensity absorption coefficient, I_m is the wave's interference-fringe amplitude distribution, and Φ is the rate of conversion of absorbed wave energy to heat. From Equation (7.8) we find that the thermal diffraction grating decays with a time constant obtained from

$$\tau_{th} = \rho C_p \Lambda^2 / 4\pi^2 K \tag{7.9}$$

where Λ is the diffraction grating spacing. This equation is obtained by replacing ∇ by k_G (the grating number) in Equation (7.8), then writing $k_G = 2\pi/\Lambda$. Furthermore, the steady-state temperature change is obtained in a similar way by $\delta T = \alpha_0 I_m \Phi \Lambda^2 / 4\pi^2 K$ from Equation (7.8). Accordingly, we get

$\delta n = (\partial n / \partial T)_\rho (\alpha_0 I_m \Phi \tau_{th} / \rho C_p)$ for the refractive-index change. The nonlinear refractive index due to these thermal effects can be expressed by

$$n_2 = (\partial n / \partial T)_\rho \, (nc\varepsilon_0 \alpha_0 \Phi \tau_{th} / \rho C_p) \quad [\text{m}^2/\text{V}^2] \tag{7.10}$$

where c is the speed of light in vacuo. The above equation's nonlinear refractive index n_2 due to heat is proportional to the time constant, showing that the response time decreases as n_2 increases and as the phase-conjugate wave reflectance increases. As we saw in Table 7.2, $(\partial n / \partial T) < 0$ for a generalized medium. The literature deals with the details of thermal effects.[9]

The electrostrictive effect often appears in liquids and solids; the change in density can be described by the following equation:[1]

$$\left(\nabla^2 - \frac{1}{v_a^2} \cdot \frac{\partial^2}{\partial t^2} + \frac{2\Gamma}{v_a^2} \cdot \frac{\partial}{\partial t} \right) \delta \rho = \frac{\varepsilon_0 \gamma_e}{2 v_a^2} (\nabla^2 |A|^2) \tag{7.11}$$

where $v_a = (\rho \beta)^{-1/2}$ is the acoustic velocity, β [m²/N] is the compression rate, $2\Gamma/v_a$ is the damping constant, $\gamma_e = 2n\rho(\partial n / \partial \rho)_T$ is the electrostriction coefficient, and A is the electric field amplitude. The above equation can be easily solved under steady-state conditions, where $\delta n = \varepsilon_0 \gamma_e^2 \beta |A|^2 / 4n$ is obtained for the refractive-index change. Consequently, we can write the nonlinear refractive index as follows:

$$n_2 = \varepsilon_0 \gamma_e^2 \beta / 2n \quad [\text{m}^2/\text{V}^2] \tag{7.12}$$

From Equation (7.1), the electrostrictive coefficient can be expressed by $\gamma_e = (1/3)(n^2 - 1)(n^2 + 2)$ for a liquid and by $\gamma_e = n^4 p_{ij}$ (p_{ij}: elasticity constant) for a crystal.[10]

(2) Particles in Solution (Radiation Pressure)

Assume a solution in which spherical particles (refractive index: n_p) are dissolved in a liquid (refractive index: n_L). If $n_p > n_L$, the spherical particles experience radiation pressure when light having a non-uniform intensity distribution propagates in the solution; the particles migrate in the direction of increasing intensity.[11] Consequently, a change in the spatial refractive-index distribution accompanies the migration of the particles. As a result, although the spherical particles themselves do not possess nonlinearity, there is still nonlinearity produced due to the radiation pressure of the light beams. Polystyrene latex is used for the spherical particles. We next explain radiation pressure in some detail.

As shown in Figure 7.1, assume a beam of high intensity at the center, propagating in the $+z$ direction, but offset from the center of the spherical particles in the liquid at injection.

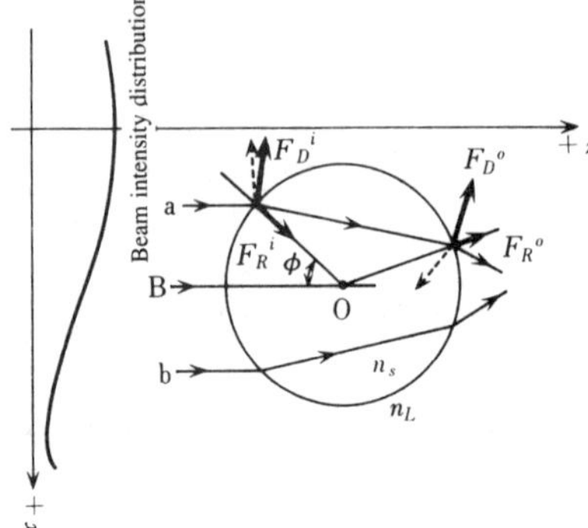

Figure 7.1 Motion of spherical particles in liquids due to radiation pressure. a and b are rays symmetric to ray B, which passes through center of sphere O.

The dashed lines are the directions of propagation of the reflected rays; n_s: refractive index of the particle; n_L: refractive index of the liquid; F_R^i, F_R^o: radiation pressure due to reflected rays on the incoming and outgoing surfaces; F_D^i, F_D^o: radiation pressure due to refracted rays on the incoming and outgoing surfaces.

The direction of pressure in the figure is for $n_s > n_L$.

Assuming a typical beam, consider two beams a and b that are the same distance off the central axis B of a sphere. The beam with high wave intensity (a) is reflected and refracted by the surface of the sphere and follows the directions indicated by the arrows. Now, since the photons also have mass, if we focus on the collisions among the particles, the particles migrate so as to satisfy the law of conservation of momentum. That is, the particles are acted on by radiation pressure. The direction in which radiation pressure is applied is almost the same as the direction of the change in the light's wave number vector since the mass of the photons, in relation to the mass of the particles, can be almost completely neglected. If subscripts R and D correspond to the contributions of reflection and refraction, the radiation pressure components on the entry and exit surfaces can be represented by F in the figure. Corresponding to the refracted intensity being greater than the reflected intensity is the fact that the magnitude of F_D is greater than that of F_R. In addition, the two reflection components cancel each other out to the first order. Therefore, the total radiation pressure exerts a force on the particles in the $-x$ and $+z$ directions. If the same concept is applied to the relatively weak beam b, forces in the $+x$ and $+z$ directions act on the particles. This also assumes that the force in the $+x$ direction due to beam b is smaller than the force in the $-x$ direction due to beam a. Consequently, the complete light beam accelerates the spherical particles away from the center of the beam and in the direction of propagation. That is to say, an optical potential corresponding to the beam's wave intensity distribution is created, and the particles are captured there. While the discussion above assumed that $n_p > n_L$, the particles drift away from regions of high wave intensity when $n_p < n_L$.

The third-order nonlinearity created by the suspended spherical particles described above can be found with the following equation:[12]

$$\chi^{(3)} = 9 m_p \left(\frac{4\pi a^3}{3}\right)^2 \frac{(\varepsilon_p - 1)\,(\varepsilon_p + 2)\,(\varepsilon_p - \varepsilon_L)\,\varepsilon_0}{k_B T} \left(\frac{\varepsilon_L}{\varepsilon_p + 2\varepsilon_L}\right)^3 \quad [\mathrm{m^2/V^2}]$$

$$(7.13)$$

where m_p is the particle concentration, a is the radius of the particles, ε_p is the relative permittivity of the particles, ε_L is the relative permittivity of the liquid, and ε_0 is the permittivity of free space. $\chi^{(3)}$ is proportional to the sixth power of the radius and inversely proportional to the temperature.

As a concrete example, $\chi^{(3)} \approx 10^{-15}$ m²/V² ($\sim 10^{-7}$ esu) for polystyrene latex spheres having diameters of 50 – 250 nm ($n_p = 1.59$) suspended in an aqueous liquid ($n_L = 1.33$).[13]

7.5 Dielectrics

(1) Origin of the Photorefractive Effect

The photorefractive effect can easily be seen in strong dielectrics. This effect was first discovered through the appearance of unusual refractive changes when $LiNbO_3$ and $LiTaO_3$ were irradiated by light.[14] The effect was subsequently verified mainly in oxides such as $BaTiO_3$, $Bi_{12}SiO_{20}$ (BSO), $Bi_{12}GeO_{20}$ (BGO) and $Sr_xBa_{1-x}Nb_2O_6$ (SBN). The magnitude of the electro-optic coefficient r_{ij} becomes particularly important, as can be understood if the refractive-index change is expressed as:

$$\delta n_i = (1/2)\, n_i^3 r_{ij} E_j \tag{7.14}$$

where the E_j are the spatial electric field components in an isotropic medium. For example, when $n_e = 2.20$, $r_{33} = 30.8$ pm/V and $E_j = 10^5$ V/m for $LiNbO_3$, $\delta n = 1.64 \times 10^{-5}$. Refer to Appendix B for the derivation of the refractive-index change for the anisotropic case.

Consider the origin of the photorefractive effect as follows. Fe impurities of the order of 1 – 10 ppm are normally contained in $LiNbO_3$ even if impurities are not intentionally added. The valence state of Fe can change from Fe^{2+} to Fe^{3+} due to oxidation of the crystal. It has been shown that the photoconductivity has a strong correlation with the Fe^{2+} concentration at the subatomic level, and that the electrons that contribute to photoconduction originate in the reaction[15]

$$Fe^{2+} + \hbar\omega = Fe^{3+} + \text{free electron} \tag{7.15}$$

That is, electrons in Fe^{2+} at the donor level acquire energy from photons and move to the conduction band to become free electrons. This process has already been illustrated in Figure 3.9. Other impurities, such as Cu (Cu^{1+}/Cu^{2+}) and Mn (Mn^{2+}/Mn^{3+}) are also understood to increase the photoconductivity. In addition, a similar process has been observed with $LiTaO_3$.

The oxygen vacancies (electric charge: 2+) create a shallow donor level in $BaTiO_3$ and the Fe creates a deep donor level.[16] The source of the donor levels contributing carriers in BSO and BGO is the color center due to trapped electrons at vacancies in Si and Ge.[15]

As a result, the carriers that are produced create an electric-current distribution caused by drift and scattering as expressed by Equation (3.57). The spatial electrostatic field that is created leads to the Pockels effect and gives rise to a refractive-index change, a mathematical explanation of which has already been dealt with in Section 3.3.

The electrical characteristics of photorefractive media can be found in Table 7.3. For convenience, the photorefractive effects for semiconductors mentioned later are also cited. We see that the electro-optic coefficient of $BaTiO_3$ ($r_{42} = r_{51}$) is unusually large. High reflectance in the production of phase-conjugate light is obtained by this effect. The relationships between typical crystal groups and electro-optic tensors are shown in Appendix B and Table B.2. Although not pointed out in Table B.2, one must be mindful of the optical activity in these media when considering polarization.

Table 7.3 Physical properties of the photorefractive effect

Parameter	Symbol	Unit	LiNbO$_3$	KNbO$_3$	BaTiO$_3$	Bi$_{12}$SiO$_{20}$	SBN	GaAs	InP	CdTe
Crystal group			3 m	4 mm	4 mm	$\overline{4}3$ m	4 mm	$\overline{4}3$ m	$\overline{4}3$ m	$\overline{4}3$ m
Wavelength	λ	μm	0.515			0.515	0.515	1.06	1.06	1.06
Refractive index	n		2.3	2.23	2.4	2.6	2.35	3.6	3.3	2.8
Static dielectric constant	ε_s		98 (ε_{11})	55	4300 (ε_{11}) 168 (ε_{33})	56	3400 (ε_3)	12.9	12.7	9.4
Electro-optic coefficient	r_{ij}	pm/V	31 (r_{33})	64 (r_{33})	1640 (r_{42}) 80 (r_{33})	5 (r_{41})	37 (r_{13}) 1340 (r_{33})	1.4 (r_{41})	1.45 (r_{41})	5.5 (r_{41})
Donor density	N_D	m^{-3}	10^{23}	—	—	10^{25}	—	—	—	—
Trap density	N_A	m^{-3}	10^{22}	—	2×10^{22}	10^{22}	—	1.5×10^{21}	2×10^{20}	$\sim 10^{21}$
Ionic cross-sectional area	S	m^2	—	—	—	1.6×10^{-23}	—	—	—	—
Coefficient of recombination	γ_R	m^3/s	10^{-15}	—	5×10^{-14}	2×10^{-17}	—	—	—	—
Transition	μ	m^2/Vs	10^{-5}	—	5×10^{-5}	3×10^{-6}	—	—	—	—
Absorption coefficient	α	cm^{-1}		—	—	0.3		1.5	1.4	—
Reference			17	18	19	19	20	21	22	23

SBN: $Sr_xBa_{1-x}Nb_2O_6$ ($x = 0.75$); $\alpha = s(N_D - N_A)$: when carriers are electrons

(2) Response Time with the Photorefractive Effect

As shown by the Equation (3.82), the generation time τ_g for a diffraction grating largely depends on the dielectric relaxation time τ_{di}, which is proportional to the static permittivity ε_s and inversely proportional to the illumination intensity I_0. Even in the same medium, increasing the optical intensity I_0 increases the electrical conductivity and shortens the response time. Consequently, the response time is dramatically shortened to $0.1 - 1$ second when I_0 is 1 W/cm^2 in BaTiO$_3$ (which has a high ε_s). As was already mentioned, τ_g also depends on the diffraction grating spacing, making order-of-magnitude fluctuations possible. The response time in BSO is in the range of 1 ms.

The response time is slow due to deep donor levels for the oxides shown in Table 7.3. Accordingly, one can (i) improve the response speed by active oxidation and reduction of oxides,[16,24] and (ii) as explained in Section 7.6(5), study the high speeds associated with semiconductors in order to increase response times. We will attempt to explain (i) through an example involving BaTiO$_3$.[16] As expressed by $\tau_{di} = \varepsilon_s\varepsilon_0/\sigma$, the dielectric relaxation time, which exerts a large influence on the total response time, is inversely proportional to the conductivity σ. As was explained above, a deep donor level is created by Fe. When the predominant carriers are holes, Fe^{3+} becomes Fe^{2+} when the medium is reduced, and the response time is shortened because of an increase in photoconductivity due to the electrons. Some means is required for reduction since Fe^{3+} is chemically more stable than Fe^{2+}.

7.6 Semiconductors

Semiconductors extend over a multiplicity of classifications, including Si and Ge of Group IV, GaAs, InSb, and InAs of Groups III-V, and CdS, CdSe, and HgTe of Groups II-VI, as well as three or four others. Consequently, there is just as large a variety of causes for nonlinear polarization. Before individual reasons for nonlinear polarization are explained, Table 7.4 lists measured data for each semiconductor type.[25] This table shows a general trend: as the nonlinearity $\chi^{(3)}$ becomes larger, there is an accompanying increase in the absorption coefficient α or in the response time τ. $\chi^{(3)}/\alpha\tau$ remains nearly constant regardless of the medium.[26] Consequently, there is a problem with τ becoming small as $\chi^{(3)}$ becomes large. That the value of $\chi^{(3)}/\alpha\tau$ is medium-independent generally applies to other naturally-existing media. Next, we explain only typical causes.

(1) Bound Electrons Under Anharmonic Potential

Bound electrons under anharmonic potential are typically a principal cause of nonlinear polarization in semiconductors. This condition is predominant when waves of sufficiently lower frequency than the band gap pass an intrinsic semiconductor. Now consider N anharmonic oscillators per unit volume. Each

Table 7.4   Nonlinear parameters of semiconductors[25]

Substance	Wave-length λ (μm)	Nonlinearity $\chi^{(3)}$ (esu)	Absorption coefficient α (cm^{-1})	Response time τ (s)	$\dfrac{\chi^{(3)}}{\alpha\tau}$	Temperature T (K)	Nonlinearity mechanism
GaAs/AlGaAs	0.84	4×10^{-2}	1.2×10^{4}	2×10^{-8}	170	300	Exciton
InSb	5.4	0.3	70	4×10^{-7}	1×10^{4}	77	Band filling
HgCdTe	10.6	5×10^{-2}	8	3×10^{-8}	2×10^{5}	77	Band filling
n–InSb	10.6	2×10^{-7}	2	4×10^{-12}	3×10^{4}	2	Non-parabolic
n–Si	10.6	3×10^{-7}	700	1×10^{-12}	430	300	Impurity
Ge	10.6	1×10^{-10}	10^{-2}	1×10^{-14}	1×10^{6}	300	Constrained electrons
p–GaAs	10.6	1×10^{-6}	5×10^{3}	5×10^{-13}	400	300	Within the valence band
HgTe	10.6	1.6×10^{-4}	3.4×10^{3}	5×10^{-12}	9×10^{3}	300	Within the valence band

oscillator's equation of motion can be written as

$$\frac{d^2x}{dt^2}+\Gamma\frac{dx}{dt}+\omega_R{}^2x+Vx^2=-\frac{e}{m^*}E\exp(i\omega t)+\text{c.c.} \tag{7.16}$$

where x is the displacement of the electron from the equilibrium position, Γ is the damping constant, ω_R is the medium's resonant frequency, Vx^2 is the anharmonicity term, e is the electron charge, m^* is the effective mass, and E is the electric field of the wave. If the contribution of anharmonicity is extremely small, Equation (7.16) can be solved by perturbation theory. Assume that a sufficiently low frequency ω ($\ll \omega_R$) of light from the band gap is injected. The following third-order nonlinearity can be obtained after the perturbation result for x is derived:[26]

$$\chi_v{}^{(3)}(\omega;\omega,\omega,-\omega)=\frac{Ne^4V^2}{\varepsilon_0 m^{*3}(\omega_R{}^2-i\Gamma\omega)^5}\quad[\text{m}^2/\text{V}^2] \tag{7.17}$$

where ε_0 is the permittivity of free space. Semiconductors to which this model applies include Ge, Si, GaAs, CdS, and CdSe, that is, those particularly in the neighborhood of the 10-μm band. The magnitude of $\chi^{(3)}$ is approximately 10^{-20} to 10^{-18} m^2/V^2 (10^{-12} to 10^{-10} esu). For CdS, $\chi^{(3)}$ is about 4.4×10^{-7} esu.

(2) Nonlinear Polarization Due to Non-Parabolic Conduction Bands

Mid-range free-carrier densities (10^{21} m^{-3}) are a cause of nonlinear polarization due to the nonlinear behavior of free carriers caused by non-parabolic conduction-band energy levels. When waves having two nearly equal frequencies (ω_f,ω_p) propagate in a semiconductor, the momentum equation for conduction-band electrons is

$$\frac{dp}{dt} + \frac{p}{\tau_m} = \frac{1}{2} e [E_f \exp(i\omega_f t) + E_p \exp(i\omega_p t) + \text{c.c.}]$$

(7.18)

where τ_m is the momentum relaxation time. Neglecting the electron energy distribution, if the resulting current density depending on conduction-band electron movement due to the injection of light is integrated with respect to time we obtain:

$$\chi_{wp}^{(3)}(\omega_4 ; \omega_f, \omega_f, -\omega_p) = \frac{Ne^4}{4\varepsilon_0 m^{*2} E_g \omega_f^2 \omega_p \omega_4} \quad [\text{m}^2/\text{V}^2]$$

(7.19)

as the third-order nonlinearity.[27] Here, m^* is the effective mass, E_g is the band-gap energy, and $\omega_4 = 2\omega_f - \omega_p$. Values in the neighborhood of $\tau_m \ll \omega_f^{-1}$ and ω_p^{-1} were used for solving the equation. In Equation (7.19), the nonlinearity is shown to become large when the band gap is narrow and the effective mass is small. These effects make a striking appearance in semiconductors with narrow band gaps such as InSb, InAs, and CdHgTe.

Additionally, when all light wave frequencies are the same, as in degenerate four-wave mixing, the effects of temperature fluctuations become important in the electron-velocity distribution. If we assume that ω_f is the pump frequency and that ω_p is the probe frequency, the resulting effect of the temperature fluctuations is:[28]

$$\chi_f^{(3)} = \chi_{wp}^{(3)} \left\{ 1 + \frac{(2\tau_{th} - \tau_m)}{\tau_m [1 - i(\omega_f - \omega_p)\tau_{th}]} \right\}$$

(7.20)

where $\chi\omega_p^{(3)}$ is obtained from Equation (7.19), and τ_{th} is the relaxation time of energy from the electron gas energy to the grating. As $\omega_f - \omega_p \to 0$ in this equation, the nonlinearity approaches a maximum. This condition has already been illustrated for *n*-InSb in Figure 6.10.

(3) Band-Filling Effect

The band-filling effect is a phenomenon whereby the excitation-state carrier density dramatically increases, saturation absorption occurs, and there is an apparent reduction in the absorption coefficient when waves having frequencies in the neighborhood of a direct (-transition) semiconductor's band-gap frequency are injected. This classical treatment was already used to explain the two-level model in Section 4.1. Called the dynamic Burstein-Moss effect, it deals with the two-band model (where there is an expansion at each level). The process is divided into the following three phases.

(i) Electrons and hole plasma are produced by incident waves

(ii) Electrons jump the conduction band, but the conduction band may not fill up because relaxation time in the band is extremely fast, thus causing saturation

(iii) Since the absorption spectrum moves toward the high-energy (short-wavelength) side due to saturation, the refractive index also changes due to the Kramers-Kronig relation.

A typical situation is shown in Figure 7.2.

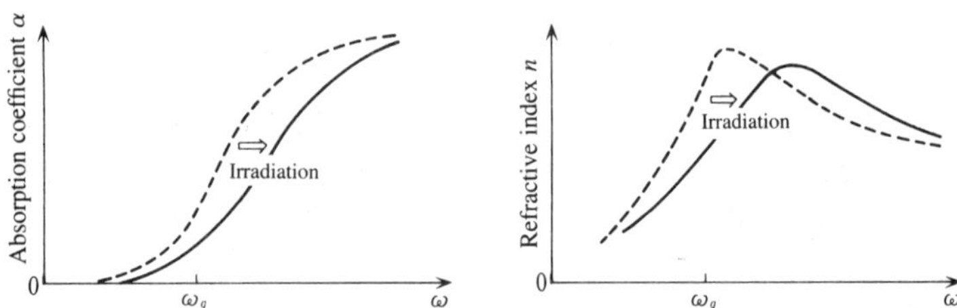

Figure 7.2 Absorption coefficient for the band-filling effect and changes due to irradiation. ω: incident frequency; ω_g: frequency corresponding to band gap.

When $\omega < \omega_g$ (ω: incident light frequency; ω_g: band-gap frequency), the nonlinear refractive index due to the band-filling effect can be approximated by:[29]

$$n_2 \approx -\frac{1}{4\varepsilon_0 n^2 c\hbar^3}\left|\frac{ep_{vc}}{m\omega}\right|^4\left(\frac{2m^*}{\hbar}\right)^{3/2}\frac{T_1}{T_2(\omega_g-\omega)^{3/2}}\quad[\mathrm{m^2/V^2}] \tag{7.21}$$

where p_{vc} is the momentum matrix element accompanying the transition from the valence band to the conduction band, T_1 is the energy relaxation time, T_2 is the phase relaxation time, and c is the speed of light. An important element of Equation (7.21) is that, for long-wavelength waves (longer than the band gap), the refractive index becomes negative and, moreover, its magnitude increases monotonically along with ω. Furthermore, if T_1 is large, n_2 becomes large because a large carrier population increase in the conduction band can be anticipated. When T_2 is small, n_2 also becomes large because the pumping rate becomes high. As shown in Table 7.4, $\chi^{(3)}$ can reach a level of approximately 1 esu (10^{-8} m²/V²).

(4) Excitons in Quantum-Well Structures

An electron and a hole bound together by Coulomb forces make up an exciton. An exciton behaves as if it were a single, neutral particle. The carrier can be restricted to one direction in a multiquantum-well (MQW) structure assembled from two types of ultra-thin semiconductor layers having thicknesses in the order of 10 nm. The effective mass of the carrier becomes extremely anisotropic, taking on a large value in a direction perpendicular to the layering. The combined energy becomes high in this type of two-dimensional exciton, and the stability of the exciton becomes possible even at ambient temperature. Characteristics of cases where excitons are used in phase-conjugate light generation include (i) a large $\chi^{(3)}$ at room temperature and (ii) low saturation intensity. Thin sample materials must be used because the coefficient of absorption is large. For GaAs/AlGaAs MQW, $\chi^{(3)} \approx 0.06$ esu (wavelength: 852 nm);[30] for GaInAs/InP MQW, with saturation light intensity $I_{sat} \approx 70$ W/cm^2, $\chi^{(3)} \approx -0.07$ esu (wavelength: 1.62 μm).[31]

(5) Photorefractive Effect

Since the photoelectric coefficient is large in the photorefractive effect in oxides, the reflectance for phase-conjugate waves is high. However, a current drawback is that response time is slow (almost a second). Accordingly, the photorefractive effect in compound semiconductors is under study, with high-speed response as the goal. GaAs, InP and CdTe are used in the 1-μm band. Fe is added to InP and Cr is added to GaAs, resulting in a comparatively low donor level, which produces a fast response time.[32] Even though the photoelectric coefficient is small in semiconductors, the change in the refractive index per unit energy is maintained at the same level as in oxides since the electrostatic dielectric coefficient is also small.[33] Table 7.3 catalogs the constants for oxides and semiconductors.

7.7 Noncrystalline Substances

(1) Dyed Glass Filters

Commercially available dyed glass filters having a cutoff in the neighborhood of 600 nm contain microscopic particles of CdS_xSe_{1-x} in silicate glass. These compounds demonstrate higher nonlinearities ($\chi^{(3)} = 1.3 \times 10^{-8}$ esu; wavelength: 532 nm); moreover, since a fast response time can be expected,[34] Corning and Schott — among other companies — are making all types of dyed glass filters, and their mechanisms are being studied.

Scanning electron micrographs show that the diameter of semiconductor particles is of the order of 10 nm.[35] These particles are used for a quantum confinement effect. (That is, electrons and holes are confined to localized deep potential wells in the base material. A degree of freedom is lost since the dimensionality is lowered, and the nonlinearity is increased.) Excitons behave similarly to electrons and holes under irradiation, and response times are short for small-diameter particles. The relaxation characteristics for a given nonlinearity and light intensity are explained by the band-filling model.[36] There are great expectations for future development of the details.

The optical damage threshold of colored glass filters is quite high at 100 MW/cm^2, and since sub-nanosecond response times have already been achieved, these devices offer significant advantages in the development of phase-conjugate optics.

(2) Glass

All types of glass exhibit a large nonlinear refractive index accompanying electron movement. The magnitude of the index is in the order of $n_2 \approx 10^{-22}$ m^2/V^2 ($\sim 10^{-13}$ esu).[37] Among these, n_2 is approximately 2.1×10^{-22} m^2/V^2 (1.9×10^{-13} esu; wavelength: 633 nm) for fused quartz (SiO$_2$). Low loss values (0.2 dB/km $\approx 4.6 \times 10^{-7}$ cm^{-1}; wavelength: 1.55 μm) have been achieved with optical fibers. Accordingly, as was learned from Equation (6.1), practical effective interaction lengths l_{eff} can be as great as 20 km.

7.8 Organic Compounds

(1) Macromolecular Single Crystals

Among organic compounds, high nonlinearity in certain types of media is a crucial concern for non-localized one-dimensional conjugate π electrons (benzene rings). Moreover, the third-order macromolecular polarizability rate γ (hyperpolarizability: the proportionality constant between the third-order local electric field and the dipole moment) for microscopic third-order nonlinearity $\chi^{(3)}$ is related to the number density of double bonds N. This was discovered in polyenes. In order to theoretically investigate this, a model with $2N$ electrons in a quantum well (infinite barrier) of length $2L$ was considered. If a perturbation calculation of the electric dipole moment due to light-electron interaction is made, the third-order hyperpolarizability for a one-dimensional system can be represented as:[38]

$$\gamma = \frac{256 L^{10}}{45 \pi^6 a_B^3 e^2 N^5} \tag{7.22}$$

where a_B is the Bohr radius and e is the electron charge. Here, if $N \propto L$, we see that $\gamma \propto N^5$ (L^5). $\chi^{(3)}$, then, increases as the chain becomes longer.

Since the basic structure of diacetylene is $R - C \equiv C - C \equiv C - R'$, various materials can be synthesized by rearranging the side chains R and R'. For example, the TCDU monomer has the form $R = R' = (CH_2)_4 OCONHC_6H_5$. This monomer can be crystalized by slow extraction of a 1:1 mixture of acetone and ethyl acetate at room temperature. Macromolecular single-crystal polydiacetylene can be synthesized by polymerization of this type of monomer through gamma-ray irradiation. Table 7.5 shows third-harmonic measured values of $\chi^{(3)}$ for polydiacetylenes.[39]

Table 7.5 Nonlinear parameters of polydiacetylene[39]

Material	Direction of polarization	Nonlinear refractive index n_0	Nonlinearity $\chi^{(3)}$ ($\times 10^{-13}$ esu)	
			$\lambda = 1.89\,(\mu\mathrm{m})$	$\lambda = 2.62\,(\mu\mathrm{m})$
TDCU monomer	$\parallel \eta$ axis	1.58	1.2	—
	$\parallel \xi$ axis	1.53	1.1	—
TDCU polymer	$\parallel$ main chain axis	1.80	700	370
	$\perp$ main chain axis	1.65	<7	<4
PTS polymer	$\parallel$ main chain axis	1.88	8500	1600
	$\perp$ main chain axis	1.58	<100	<20

Characteristic of this are the striking increase in the value of $\chi^{(3)}$ in polymers and the strong dependence of the polarization on the orientation of the main chain. This shows that $\chi^{(3)}$ is augmented by π electrons localized along the main chain. The maximum nonlinearity is $\chi^{(3)} \approx 8.5 \times 10^{-10}$ esu (wavelength: 1.89 μm) for PTS (polydiacetylene toluene sulfonamide) polymers. Since 1.89 μm is closer than 2.62 μm is to the absorption peak in the case of PTS polymers, the large $\chi^{(3)}$ value at 1.89 μm is believed to be due to resonance effects. Polydiacetylenes become transparent at wavelengths of 1 to 2 μm.

Polydiacetylenes are used not only as crystals, but also as thin films.[40]

(2) Dyes

Saturation absorption plays a very important role when dyes are used for the production of phase-conjugate light. We will consider the three-level model shown in Figure 7.3 in order to explain this.[41] When light is injected, the absorbed energy excites electrons from the S_0 level to the S_1 level. The majority of the electrons phosphoresce and then return to the S_0 level after a T_1-level lifetime τ, due to the nonradiative transition as electrons move from the S_1 level to the T_1 level. The lifetime determines the response speed of the system. On the other hand, some of the electrons at the S_1 level produce fluorescence and likewise return to the S_0 level. Among the electrons making the transition from the S_1 level to some other level, the percentage of electrons going to the T_1 level is called the triplet yield and is represented by Q. When Q is near 1, the number

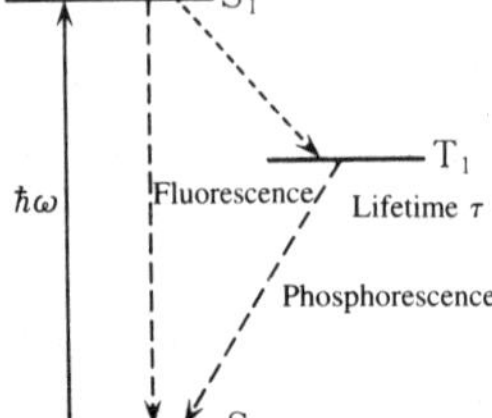

Figure 7.3 Three-level model for organic dyes.

S_0, S_1: Singlet level
T_1:   Triplet level
τ: Level T_1 lifetime

of electrons at the S_0 level is reduced only by the number of electrons occupying the T_1 level. Therefore, the number of excitable electrons at the S_0 level is dramatically reduced if the intensity of the injected waves is high at a frequency corresponding to the S_0- and S_1-level energy; the rate of absorption is thus commensurately reduced. This phenomenon is called saturation absorption. The product of the saturation intensity I^0_{sat} and the response time τ is expressed by the equation

$$I^0_{sat}\tau = \hbar\omega/\sigma Q \tag{7.23}$$

where σ is the absorption cross-sectional. Since σ is usually regarded as a constant in frequently-used dyes, a decrease in saturation intensity leads to an increase in response time. The value of the product $I^0_{sat}\tau$ is at the 10^5 to 10^6 [Wcm^{-2} nsec] level.[4] The response time is in the order of milliseconds to nanoseconds. The dyes can be dissolved in ordinary organic solvents. The above-mentioned product is relatively small for rhodamine 6G/cresol violet in ethanol and β-carotene in hexane.

8.
Applications of Phase-Conjugate Light

As explained in the chapters up to now, phase-conjugate light has a variety of characteristics and properties not associated with normal light. Consequently, we can attempt applications in a variety of areas that make use of those features. We will present in this chapter a general overview of applications, followed by an explanation of individual application examples related to the features peculiar to phase-conjugate light.

8.1 Introduction

The characteristics of phase-conjugate waves can be enumerated as follows:

(i) Waves that pass through an aberrating medium are injected into a phase-conjugate mirror. If the outgoing wave is repropagated through the same medium, the wave's phase distortion is compensated. (Phase-compensation effect)

(ii) Since phase-conjugate waves are produced by interaction in a nonlinear medium, the spatial multiplicative effect among the light waves appears in the phase-conjugate wave. (Space-domain multiplicative interaction)

(iii) The phase shift of the electric field depends on the wave intensity, due to the optical Kerr effect in the nonlinear medium. (Wave intensity dependency of phase shifts)

(iv) The reflectance strongly depends on the detuning between the pump and probe wave frequencies. In addition, the phase-conjugate wave also reflects frequency dependencies inherent in the medium. (Detuning dependency)

(v) A phase-conjugate wave can be regarded as a time-inverted wave since its direction of propagation is exactly opposite that of the probe wave and the wave front is identical to that of the probe wave. (Time inversion)

(vi) When pulsed waves are used in the production of phase-conjugate light, the resulting polarization, that is, the generated phase-conjugate wave intensity, depends on the product of the interacting waves in the waves' temporal domains. (Time-domain multiplicative interaction)

(vii) When the waves are treated quantum-theoretically, the probe and phase-conjugate waves correspond to the photon annihilation and creation operators, respectively, and a correlation exists between them. (Quantum correlation)

Classifications of the features and application areas of phase-conjugate light are given in Table 8.1. Even though some theoretical proposals are at a standstill, some are actually being implemented. Overall, however, many are awaiting new development.

Table 8.1 Phase-conjugate wave characteristics and application areas

Phase-conjugate wave characteristic	Application areas		
		Data control	Manufacturing/ power generation	Other
Phase correction activities	Optical interferometer Optical gyroscope Optical image transmission Satellite communications		Laser resonator Laser fusion Photolithography	Automatic tracking
Multiplicative interactions (space-domain)		Optical calculators Pattern recognition Matched filters Optical memory		
Phase-shift wave intensity dependence		Optical bistability		
Detuning dependence	Optical frequency filters Semiconductor laser amplifier			Nonlinear polarization
Time inversion	Group delay equalizer			
Multiplicative interactions (time-domain)				Wave pulse compression
Quantum correlation	Optically squeezed state			

8.2 Applications of Phase-Correction Functions

The phase-change compensation that was discussed in Section 2.2 can only be achieved when the phase is reciprocal and is quasistatic, changing very little during the round trip along the wave path. Consequently, phase changes in round-trip wave paths cannot be offset in non-reciprocal phenomena such as (i) changes in time dependence, (ii) the Faraday effect, and (iii) the Sagnac effect. These types of features are also being used in applications.

(1) Optical Resonators

The length of a conventional resonator changes during operation because of thermal effects. There are also cases of mode hopping and frequency drift. When a solid-state laser in an active medium is excited by a flash lamp, the light beams in the medium suffer aberrational effects because of the laser medium's thermal expansion and the temperature dependence of the refractive index. In addition, the nonuniform temperature distribution produces complex refraction via thermal stress which interferes with the polarization.[1] This type of problem is solved by utilizing a phase-conjugate mirror as one of the wave resonator's mirrors to compensate the polarization or to reduce the wave's beam width to approach the diffraction limit. This type of resonator is called a phase-conjugate optical resonator, a schematic diagram of which is shown in Figure 8.1. As described by the basic characteristics of phase-conjugate mirrors in Chapter 2, phase-distortion correction is automatically performed during round trips of a light beam even if there are aberrating substances within the optical resonator.

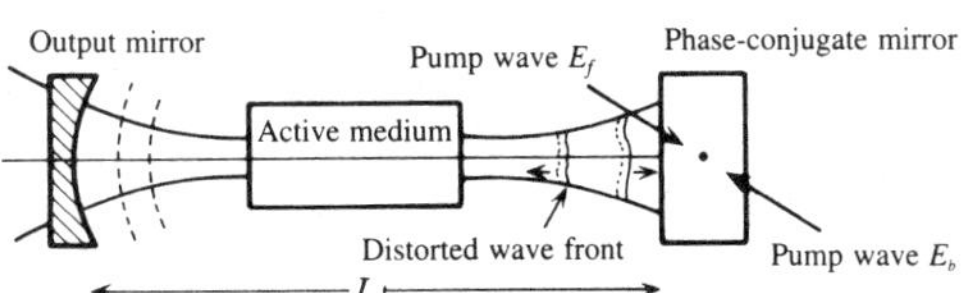

Figure 8.1 Application of a phase-conjugate mirror in a resonator.

The following are advantages of phase-conjugate optical resonators.

(i) Light beams near the diffraction limit can be extracted because phase distortion (aberration) caused by thermal distortions and multiple diffraction produced in the active medium can be offset by the light beam's round trip in the resonator.

(ii) The longitudinal mode of the wave resonator does not depend on the length of the resonator. Consequently, the laser resonator oscillates at a wavelength within the amplification band of the medium and the phase-conjugate wave generating medium; oscillation continues even if the length of the resonator is changed.[2] The reasons for this are as follows. The longitudinal mode establishes the phase difference for a light beam's round trip in a resonator as an integral multiple of 2π. When a phase-conjugate mirror is used, the phase difference along the propagation path becomes zero in one round trip due to the phase-correction effect. However, strictly speaking, as shown by Equation (2.4), since reflection accompanies the phase difference ϕ_{pc}, two round trips are required to offset the effects of ϕ_{pc}. This characteristic is in sharp contrast to the occurrence of mode hopping and frequency drift in conventional laser resonators when the length of the resonator is changed due to heat. That is, the longitudinal mode spacing of a conventional laser resonator is $c/2L$ (c: velocity of light; L: resonator length).

(iii) Characteristic (ii) shows that a phase-conjugate resonator can be used to synchronize with the master laser frequency and to continuously vary the frequency.

(iv) The wave resonator's transverse mode has the same frequency.
We next look at resonator behavior via a ray matrix. If r_i and r'_i are the ray's height above the axis and inclination, respectively, then the ray's propagation can be described by the ABCD matrix

$$\begin{bmatrix} r_{\text{out}} \\ r'_{\text{out}} \end{bmatrix} = \begin{bmatrix} A & B \\ C & D \end{bmatrix} \begin{bmatrix} r_{\text{in}} \\ r'_{\text{in}} \end{bmatrix} \tag{8.1}$$

Therefore, the ray matrix that shows the ideal phase-conjugate mirror's reflective characteristic can be expressed by

$$\begin{bmatrix} A & B \\ C & D \end{bmatrix} = \begin{bmatrix} 1 & 0 \\ 0 & -1 \end{bmatrix} \tag{8.2}$$

If we consider a Gaussian beam, this can be written as

$$\frac{1}{q} = \frac{1}{R} - \frac{i\lambda}{\pi w^2} \tag{8.3}$$

where q is the complex radius of curvature, R is the radius of curvature, w is the beam-spot size, and λ is the wavelength. The relationship between the radii of curvature of the wave incident on the phase-conjugate mirror (q_i) and the reflected wave (q_r) can be expressed by the following equation:[3]

$$q_r = \frac{A q_i^* + B}{C q_i^* + D} \tag{8.4}$$

Consequently, $q_r = -q_i^*$ for the case of an ideal phase-conjugate mirror; if there are two round trips the arbitrary electric field is precisely restored to the original electric field independent of the radius of curvature of the concave mirror.

Amplification is possible with a phase-conjugate mirror using four-wave mixing. Accordingly, if such a mirror is used as one of the mirrors in an optical resonator, the laser may oscillate even if there is no medium for amplification inside the resonator. An example of this type of phase-conjugate optical resonator is shown in Figure 8.2.[4] A resonator is formed between a $BaTiO_3$ crystal and a spatula, and oscillation continues independent of the direction of the spatula.

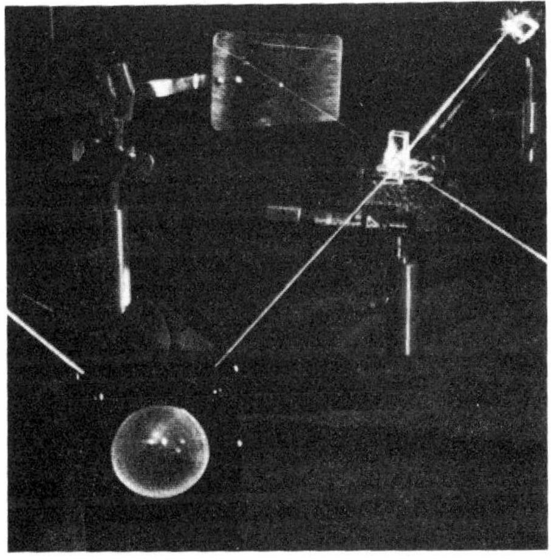

Figure 8.2 Phase-conjugate wave resonator laser using a cooking spatula. *(After Feinberg and Hellwarth)*[4]

The pump wave from the lower left of the figure is injected into a $BaTiO_3$ crystal and its transmitted wave is reflected by the mirror at the upper right in this counterpropagating setup. If the ray reflected from the spatula's surface (center top) faces the crystal at this time, a resonator is formed by the spatula and the crystal and the laser oscillates even though there is no active medium. Oscillation continues even if the direction of the spatula is changed.

(2) Auto-Targeting

It is very important in laser nuclear fusion and optical lithography that a narrow light beam be concentrated in a very small region. However, as was explained for the case of an optical resonator, the light beam tends to spread due to a variety of causes such as distortion of the amplifying medium and optical-system aberrations. Consequently, we will consider the use of phase-conjugate mirrors in what follows.

Figure 8.3 shows schematically an optical system for concentrating light beams into small regions.[5] In (a), the reference wave is injected from the side and an optical focusing system (e.g., a lens) concentrates it on a spherical target.

Scattered and reflected light from the target becomes spherical waves, some of which return to the lens. These waves become parallel after passing through the lens and enter the optical system to the right. If we also assume an aberrating substance here, phase correction is performed while the light beam makes a round trip through the beam splitter and phase-conjugate mirror. This is a basic outline of how to concentrate light on a target. However, in general, obtaining an intact spherical wave from an actual target is difficult; there is also a risk of the target being damaged by the reference wave. Consequently, there are cases where the target and the optical reference are separated.

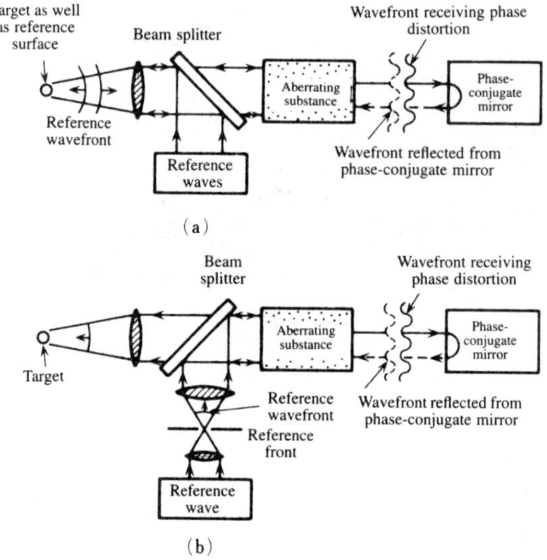

Figure 8.3 Automatic target effect of a phase-conjugate mirror.

(a) when the target doubles as the reference surface
(b) when the target and the reference surface are separate

Figure 8.3(b) is an example of a system in which the reference optics are separated from the target.[5] The reference section involves a beam splitter (BS) which is symmetrically positioned with respect to the target. The optical system's lenses are also symmetric with respect to BS. The reference wave is injected from the side under these conditions with the light first being guided by BS to the optical system at the right. Even if there is an aberrating substance in the propagation path, the light beam is phase-corrected between the beam splitter and the phase-conjugate mirror during its round trip, and is concentrated on the target. In this configuration the beam splitter and the lens of the optical system must be adjusted so that the spherical wave reflected by an actual target corresponds precisely with the reference spherical wave injected from the side.

a. Laser Fusion

A light beam is concentrated into a small surface area in laser fusion, increasing the optical energy concentration and approximating the conditions that occur in fusion. In order for the energy concentration to be increased, it is necessary that the amplifying medium be connected in multiple stages, and that beams from several directions be focused on the spherical target (a fuel pellet). The spherical target has a diameter of several hundred microns and is filled with heavy hydrogen (deuterium). The conventional way of adjusting and concentrating a complex light beam requires highly sophisticated techniques such as using a sensing servo mechanism. Use of phase-conjugate mirrors in laser fusion is being considered to solve this type of problem. A schematic diagram of a multiplexed beam irradiation system is shown in Figure 8.4.[5]

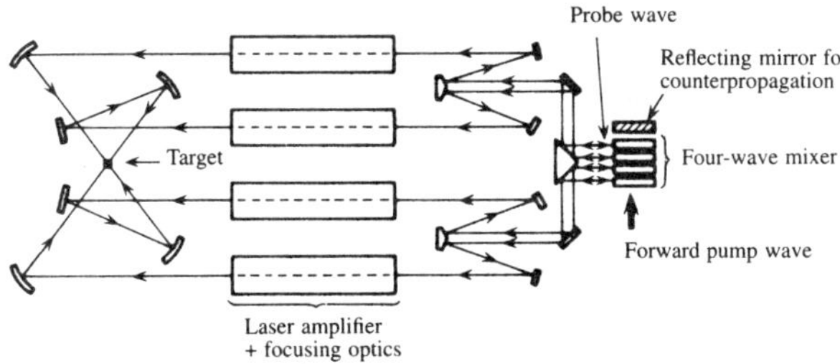

Figure 8.4 Schematic of multiple-beam, laser-fusion optical system. *(After O'Meara)*[5]

First, a light pulse having a wavelength within the amplification band of the active medium irradiates the target. The setup comprises many optical systems and several laser amplifiers placed between the target and the phase-conjugate mirrors. The light reflected from the target becomes the reference wave, and enters the system via the optical focusing system. There is an accumulation of phase distortion since there are many elements in the optical system. The reference wave amplified by the active medium becomes the probe wave and is injected into the phase-conjugate mirror. The phase-conjugate light producing system in the figure has a counterpropagating format. After reflection by the phase-conjugate mirror, the wave is again amplified and the phase distortion mentioned above is cancelled. Then, following a repetition (of amplification and phase-distortion cancellation), the energy stored in the laser amplifier is extracted in the form of a narrow light beam that is near the diffraction limit.

b. Photolithography

To increase the level of integration of components in photolithography, it is necessary to draw minute resist patterns with narrow light beams on the resist. Since the mask adheres to the resist, the light diffracts into the substrate and degrades the spatial resolution. A complex optical arrangement with a small F number is usually used to produce a small, bright focal spot with a light beam near the limit of diffraction. In principle, an optical system similar to the one used for fusion can also be used for photolithography.

An actual example is shown in Figure 8.5.[6]

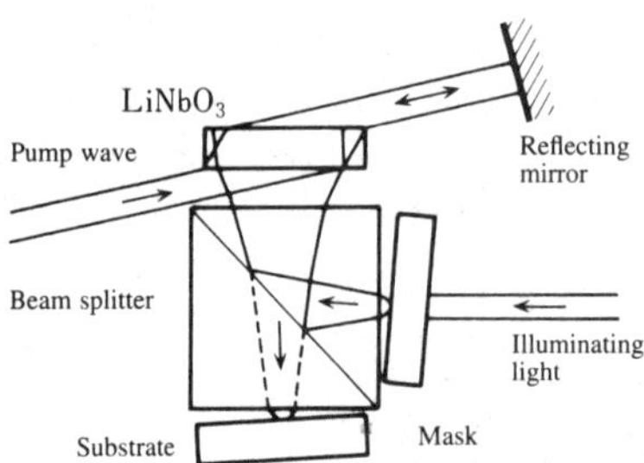

Figure 8.5 Schematic diagram of a photolithography system. *(After Levenson, Johnson, Hanchett, and Chiang)*[6]

The mask that draws the minute pattern is separated from the resist substrate and does not come in contact with it. The light source is a Kr ion laser (wavelength: 413 nm) with the pump light and illuminating beams kept separated. The pump waves are expanded in varying magnification in the horizontal and perpendicular directions, and become parallel after reaching beam diameters of 2.5 mm and 7.5 mm, respectively. The illuminating waves that penetrate the mask are reflected by a beam splitter and are injected into the phase-conjugate mirror. The phase-conjugate light producing medium is 25 mm × 25 mm LiNbO₃ used in a counterpropagating configuration. The optical-system aberrations that exist between the beam splitter and the phase-conjugate mirror are offset by the round trip the light beam makes and the reduced spot of light is concentrated on the resist. Resolution of the order of 0.55 μm has been achieved.

c. Optical Communications with Satellites; and Power Transmission

A light beam at the diffraction limit is effective when used over long distances. Optical communications with satellites and power transmission have been considered as examples of this.[2] However, this is only in the conceptual stage.

d. Auto-Tracking

The same principle can apply to the movement of substances. Automatic tracking of moving bodies has become possible. The requirement here is that it take less time for a body to move than the time required for a round trip of a light beam.

(3) Optical Image Transmission
a. Characteristics of Optical-Fiber Image Transmission

Since multi-mode optical fiber can two-dimensionally propagate several modes at one time, it can be used as a transmission path for optical images. Taking the center of the optical fiber's core as the origin, it can be described by the cylindrical coordinate system (r, θ, z). If we assume that z is the axis of light propagation, β_m is the propagation constant for each mode among the optical fibers (distinguished by the index m) and $\Psi_m(r,\theta)$ is an eigenfunction, then the following equation can represent the electric field distribution after propagation over a distance z:

$$E(r, \theta, z) = \sum_m A_m \Psi_m(r, \theta) \exp(-i\beta_m z)$$

(8.5)

where the A_m are the mode expansion coefficients. Now here the stipulation for an undistorted injected optical image after propagation over a distance z_p can be given as

$$\beta_m z_p = 2\pi \cdot (\text{an integer}) + (\text{a constant})$$

(8.6)

A very simple example that satisfies Equation (8.6) is a case where the mode propagation constants are equally spaced. For example, when $\beta_m = \beta_0 - m\,\beta_l$ (β_0 and β_l are constants), l is an integer, and $z_p = l(2\pi/\beta_l)$, we obtain a non-distorting distortion-free image for each distance $2\pi/\beta_l$. An example that satisfied the above-mentioned stipulation was discovered with meridional rays in step-index fibers.[7] When the refractive-index distribution is given by

$$n(r) = n_o \operatorname{sech}(gr)$$

(8.7)

the propagation constant becomes

$$\beta_m = n_o k - g[m + (1/2)]$$

(8.8)

where n_o is the refractive index at the core center, $g = (2\Delta)^{1/2}/a$ is the focusing constant, k is the in-vacuo wave number, Δ is the relative refractive-index difference, and a is the core radius. In practice, implementing the refractive-index distribution shown in Equation (8.7) is difficult.

The following second-order distributions exist for optical fibers, and are often used in applications:

$$n^2(r) = \begin{cases} n_o^2[1 - (gr)^2] & ; \ r \leq a \\ n_o^2[1 - (ga)^2] \equiv n_{cl}^2 & ; \ r > a \end{cases}$$

(8.9)

Using the WKB rule, we obtain

$$\beta_m = n_o k [1 - (2\Delta m / M)]^{1/2} \tag{8.10}$$

as the propagation constant for this optical fiber, where $m = 2\mu + \nu + 1$ is the number of principal modes, ν is the azimuthal mode degree, μ is the radial mode degree, and $M = m_{\max} = n_o ka(2\Delta)^{1/2}/2$ is the maximum number of modes. Although Equation (8.10) clearly does not satisfy the above-mentioned stipulations for non-distortion, it does approximately fulfill the non-distortion condition when Δ is very small. If the image degradation distance z_s due to differences in the mode propagation constants is defined by

$$(\beta_m - \beta_0) z_s \ll \pi \tag{8.11}$$

then we obtain the following equation (keeping small terms up to second order):[8]

$$z_s \leq [(2\pi)^2 n_o]/(\lambda g^2 M^2) = \lambda/(n_o \Delta^2) \tag{8.12}$$

As an example, if we assume a quartz optical fiber as the propagation path, when the wavelength $\lambda = 1$ μm, $\Delta = 0.5\%$, and $n_o = 1.45$, we obtain $z_s = 2.8$ cm, and we see that some concept other than optical fiber is necessary to perform optical image transmission.

b. Optical Image Transmission Using Phase-Conjugate Light

Figure 8.6 shows the principle of optical image transmission when multimode optical fiber is used for the transmission path.[9] Assume that optical data Ψ_m is carried by a propagating wave of frequency ω. If this data propagates through a multi-mode optical fiber for a distance z, the phase of the modes shifts by no more than $\beta_m z$, thus distorting the optical image because the propagation constant of each mode changes at the exit end of the optical fiber. If this image is injected into a phase-conjugate mirror made from a nonlinear medium, phase-conjugate light is output from the medium. This light wave offsets the phase shift of each mode if it is propagated through a multi-mode optical fiber having the same characteristics and the same length as the one mentioned above; an optical image having no distortion is obtained at the output end.

An experimental example based on this principle is shown in Figure 8.7.[10] A continuous-wave simple, longitudinal-mode Kr ion laser (wavelength: 647 nm; coherence length: 10 m) is used as the light source and a BaTiO$_3$ crystal is used as the nonlinear medium. Phase-conjugate light was produced by degenerate four-wave mixing. The laser light was separated into pump and probe waves by BS$_1$ and was further separated into two pump waves by BS$_3$. A counterpropagating configuration was used. Typical pump wave powers were $I_f = 3.2$ mW and $I_b = 7.1$ mW. The angle formed by the forward pump wave and

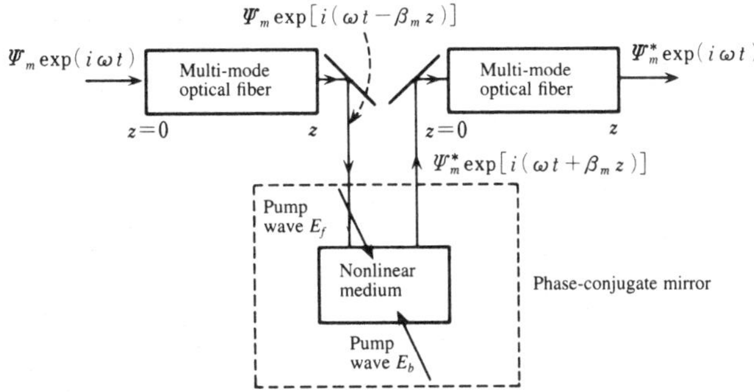

Figure 8.6 Principle of optical image transmission using a multi-mode optical fiber phase compensator depending on phase-conjugate light.

the probe wave in the crystal was 7° and the diffraction grating vector formed an angle of 18° with the crystal's c axis. The transmission path was a step-index, multi-mode optical fiber (core diameter: 85 μm; length 1.75 m) and the objective lenses (L_3 and L_4) used for coupling the laser beam to the optical fiber had a magnification of 7 (*NA*: 0.2). Phase distortion was corrected here by a round trip through the same optical fiber. Figure 8.7(b) shows an original photograph of a resolution test chart; Figure 8.7(c) is the output image when there is only one-way propagation; and Figure 8.7(d) is the reconstructed image. The spatial resolution in this example was 40 lines/mm. Correction of phase distortion is possible even for round trips through optical fibers, suggesting that mode changes have a small influence and that the optical fiber may be exhibiting some kind of reciprocal effect.

(4) Interferometers

If phase-conjugate light is applied to an interferometer, (i) the phase change doubles and the sensitivity increases and (ii) the interference fringes are visible independent of changes in the intensity of the incident light.

A schematic of a phase-conjugate interferometer is shown in Figure 8.8. A counterpropagating configuration is used to generate the phase-conjugate waves. Although the basic format is that of a Mach-Zehnder interferometer, the attitude of beam splitter BS_1 is rotated 90 degrees when compared to a conventional interferometer in order to reflect phase-conjugate waves. The image of the signal wave E_s is formed on BS_1 by lens L_1. The phase-conjugate waves produced in the nonlinear medium also form an image on BS_1 as a result of phase correction.

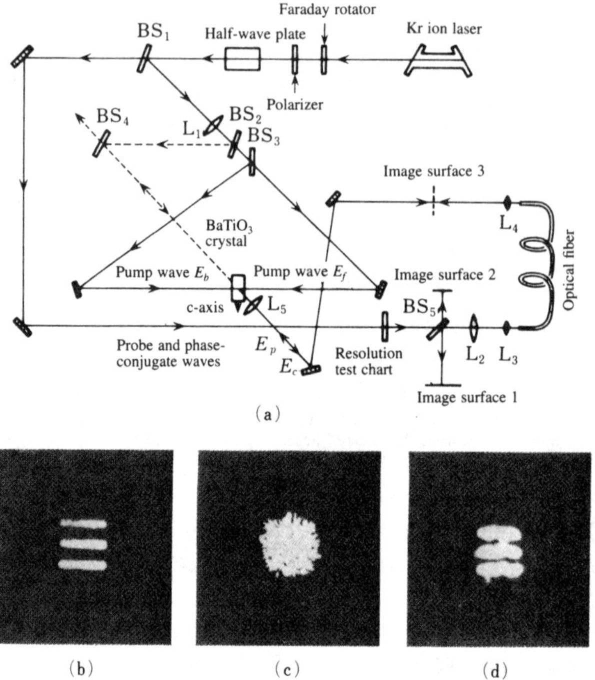

Figure 8.7 Optical image transmission using multi-mode optical fiber.
(After Dunning and Lind)[10]

(a) Experimental setup using phase-conjugate light,
(b) input image, (c) image after one-way transmission by optical fiber,
(d) reconstructed image after round trip through optical fiber.

Consequently, the length of the transmitted signal wave E_t along the path from the signal wave to BS_2 is equivalent to that of the phase-conjugate wave E_c. In addition, as shown by the diagrams of the wavefront, the phase-conjugate wave is a time-reversed wave where the phase difference between E_t and E_c at BS_2 is twice that of the phase $\phi_s(x,y)$ of the injected signal wave.

When the complex amplitude distribution of the signal wave injected at BS_1 is assumed to be $E_s(x,y) = |A_s| \exp(i\phi_s)$, the amplitude of the observable wavefront of the signal wave and the phase-conjugate wave are given by $E_t = E_s \sec(L/l_0)/\sqrt{6}$ and $E_c = -E_s{}^* \tan(L/l_0)/\sqrt{6}$, respectively, where l_0 is twice the characteristic attenuation length of the nonlinear medium, and is directly proportional to the pump-wave intensity. The interference pattern is given here as follows:[11]

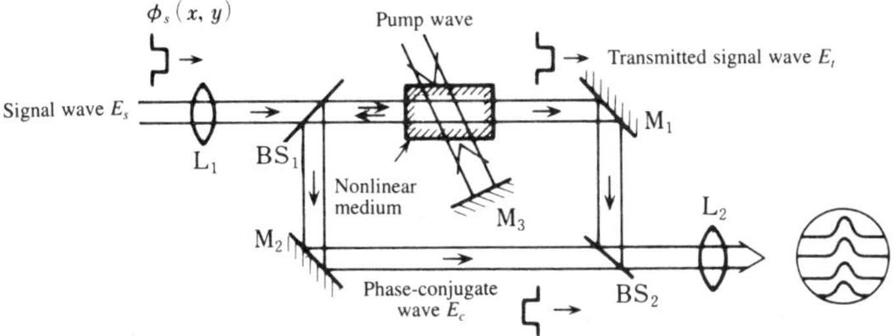

Figure 8.8 Schematic diagram of interferometer using phase-conjugate light. *(After Hopf)*[11]

The convex symbol schematicizes the change in the wave front. $\phi_s(x,y)$ is the spatial phase distribution.
Lens L_1: imaging lens for BS_1
Lens L_2: imaging lense for BS_1 observation plane
BS_1: 50% beam splitter
BS_2: 33% reflective beam splitter

$$I \propto |E_s(x, y)|^2 \sec^2(L/l_0)$$
$$\cdot \{[2 - \cos^2(L/l_0)] - [2\sin(L/l_0)]\cos(2\phi_s)\} \tag{8.13}$$

We see from this that when the gain is high, $L/l_0 \approx \pi/2$, and the fringe contrast $v = (I_{max} - I_{min})/(I_{max} + I_{min})$ becomes relatively insensitive to changes in l_0 related to pump-wave fluctuations. The same results are obtained if one of the reflective mirrors in a Michelson interferometer is replaced by a phase-conjugate mirror.

The phase-conjugate interferometer employed an argon ion laser (wavelength: 514 nm) as the light source and eosin (a saturable absorptive organic dye) dissolved in a thin film of keratin as the nonlinear medium to corroborate results for continuous and pulsed waves.[12] Thin films of 30 to 80 μm were used and a reflectivity of 10^{-3} was obtained.

If the wavefront memory time in the phase-conjugate mirror was sufficiently long, changes in shape could be observed according to the shifts in the line patterns of the interference fringes when varying intervals were assumed for the phase-conjugate wavefront interference.[13] This was attempted by using BSO crystals and organic dyes.

(5) Optical Gyroscopes

Figure 8.9 is a schematic diagram of a phase-conjugate wave gyroscope using a Michelson interferometer.[14] An optical fiber loop is inserted in one leg of the light path, and the reflective mirror in the light path is replaced by a phase-conjugate mirror. The phase change for the outward journey from point A to point B is given by

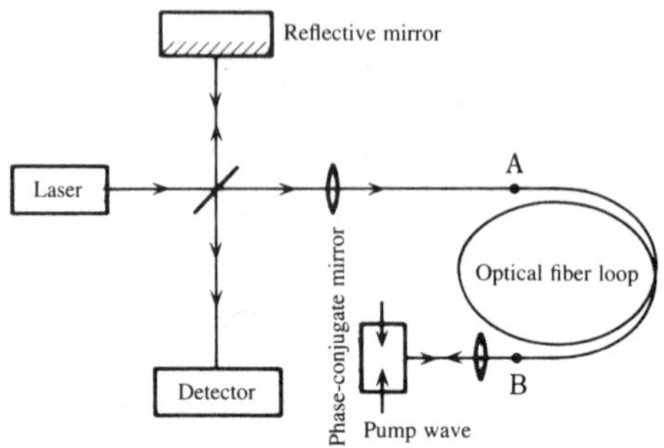

Figure 8.9 Phase-conjugate wave gyroscope using a Michelson
interferometer. *(After Yeh, McMichael and Khoshnevisan)*[14]

$$\phi_f = k_m L - (2\pi R L \Omega_R / c\lambda) \tag{8.14}$$

where k_m is the wave number inside the medium, L is the length of the optical
fiber, R is the radius of the optical fiber coil, Ω_R is the rotation rate, c is the
speed of light, and λ is the wavelength. The second term in Equation (8.14) is
the so-called Sagnac effect, which depends on rotation. The phase change in the
return path can be obtained from

$$\phi_b = k_m L + (2\pi R L \Omega_R / c\lambda) \tag{8.15}$$

Since the direction of propagation in the rotation-dependent term is opposite to
the rotational direction, the expression is reversed compared to the outgoing
path. Now, if the phase-conjugate mirror is injected with light transmitted
through the optical fiber loop, since the outgoing wave in the expression is
opposite to the phase shown in Equation (8.14), we obtain

$$\phi = 4\pi R L \Omega_R / c\lambda \tag{8.16}$$

for the total phase change after a round trip through the optical fiber loop. This
phase change can be measured by the interference fringes created with the reference wave in the other leg. Since the phase change in Equation (8.16) is
proportional to the rotation rate Ω_R, the rotation rate can be measured directly
from the phase change without eliminating the effect of the reciprocal term $k_m L$.

This procedure was used in preliminary experiments with an argon ion laser
(wavelength: 514 nm) as the light source, $BaTiO_3$ as the nonlinear medium, and
approximately 7 m of 10-cm-diameter optical fiber. Although the reflectance of
the phase-conjugate mirror was 50% and the response time was 0.1 s, the rotation rate could be measured in seconds.

8.3 Applications of Space-Domain Multiplicative Interactions

As explained in Section 3.4(2), since four-wave mixing is intimately related to holography, it can be applied to those areas where holography is already being used. Phase-conjugate light has the advantage that it can make use of amplification by a nonlinear medium.

(1) Optical Processors

Consider the optical setup shown in Figure 8.10.[15]

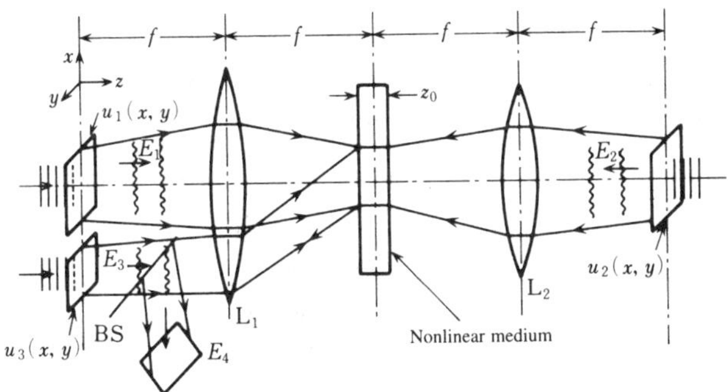

Figure 8.10 Optical operations using space-domain multiplicative operations for producing phase-conjugate light. *(After Pepper, Au Yeung, Fekete, and Yariv)*[15]

E_j ($j = 1 - 3$): incident plane waves; $u_j(x,y)$: amplitude transmittivity; E_4: outgoing wave when distance from lens L_1 is f; f: focal length of lenses L_1 and L_2.

Three plane waves E_j having frequency ω and amplitudes A_j ($j = 1$ to 3) irradiate a transmissive medium from the left and right. Assume that an image is formed in a nonlinear medium (nonlinearity $\chi^{(3)}$) by a convex lens having focal length f. If we further assume that the amplitude transmittances of the medium are $u_j(x,y)$, then the complex amplitude at the time each wave reaches the nonlinear medium is proportional to the transmittance. Since each lens implements a Fourier transform, the outgoing waves also have frequency ω when the nonlinear polarization created by the three waves is

$$P_{NL} \propto \chi^{(3)} \tilde{u}_1 \tilde{u}_2 \tilde{u}_3{}^* \tag{8.17}$$

where the tilde (~) denotes the Fourier transform. If $\chi^{(3)}$ is spatially uniform, the complex amplitude of the light wave produced by this polarization on the focal plane made by lens L_1 and beam splitter BS becomes

$$A_4 \propto \tilde{\chi}^{(3)} * u_1 * u_2 \star u_3{}^* \tag{8.18}$$

where ✻ and ★ represent the convolution and correlation operators, respectively. These operations are made possible by using Equation (8.18). As an example, if u_3 is taken as the point of origin, u_1 and u_2 become the convolution operators. If u_1 (u_2) is the point of origin, then u_2 (u_1) and u_3 become the correlation operators.

These operational functions are being applied to holography for spatially matched filters, pattern recognition and contour emphasis.

(2) Associative Optical Memory

For the sake of simplicity, assume that the desired image amplitude is $A(u,v)$ and that the amplitude of a reference wave is $B(u,v)$. Assume further that this optical data is preserved as a hologram. If a light wave of a distorted or incomplete image of A having complex amplitude $A_d(u,v)$ irradiates the hologram, then the transmitted amplitude can be obtained by

$$A_d|A+B|^2 = A_d(|A|^2 + |B|^2) + A_dAB^* + A_dA^*B \tag{8.19}$$

The last term in this equation shows the convolution operator in relation to the correlation operator B of A_d and A^*. Therefore, when A_d and A^* are equal or nearly equal, their correlation develops a sharp peak and B is reproduced. If the last term of Equation (8.19) irradiates the hologram one more time in order to extract the desired image, the following equation is obtained:[16]

$$A_dA^*B|A+B|^2 = A_dA^*B(|A|^2 + |B|^2)$$
$$+ (A_dA^*)|B|^2A + A_dA^{*2}B^2 \tag{8.20}$$

Since most substances have a uniform optical-intensity distribution, the reproduced image of A is obtained from the second term on the right-hand side of Equation (8.20). A proposal somewhat similar to this has been suggested.[17]

Figure 8.11 is an example of an optically associative memory configuration based on the above-mentioned principle.[16] The angle of entry of the reference wave was varied and images containing four patterns were recorded in a hologram by multiple exposure. A phase-conjugate mirror was created by counterpropagating degenerate four-wave mixing for which the light source was an argon ion laser (wavelength: 514 nm; pump-wave intensity of the order of 10 W/cm²), and the nonlinear medium was $BaTiO_3$. If the substance wave of the incomplete-injected image containing one part of the holographically stored data irradiated the hologram, the light was diffracted in the relatively strong direction. This diffracted wave became the probe wave and was injected into the nonlinear medium. The wave exiting the nonlinear medium propagated in a direction opposite the reference wave. If this outgoing wave irradiated the hologram again, the diffracted light became the reconstructed image.

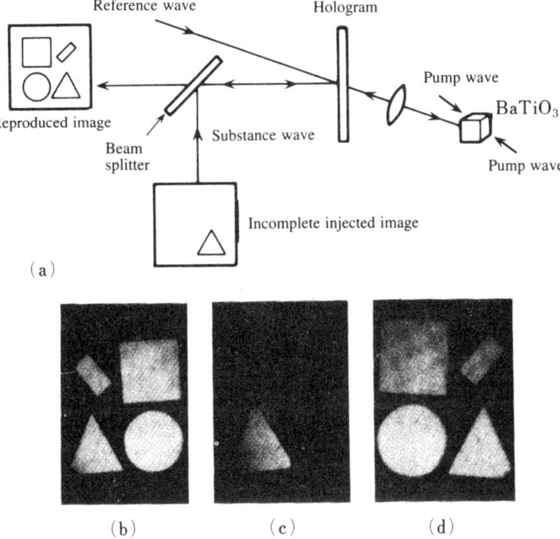

Figure 8.11 Optical memory using a phase-conjugate mirror. *(After Soffer, Dunning, Owechko, and Marom)*[16]

(a) Optical system: Original image is stored beforehand in a hologram. If the incomplete injected image containing one part of its data illuminates the hologram, the image is played back; (b) image stored as a hologram; (c) incomplete injected image; (d) image played back via optical storage.

There exists a report containing detailed commentary on associative optical memory.[18]

8.4 Using the Relationship Between Phase Shift and Wave Intensity (Bistability)

If a nonlinear medium is used to produce phase-conjugate light, the phase of each light wave changes due to the optical Kerr effect. The precise expression for phase-conjugate light production in an optical Kerr medium was given by Equation (3.20). The first term on the right-hand side of that equation shows the change in wave intensity as a function of each wave's phase. We get a hint of the use of optical bistability if we employ changes in pump-wave intensity. Characteristics with respect to pump-wave depletion were analyzed in Section 3.2(4) and could be explained in terms of elliptical functions. The coupled intensity dependence of phase-conjugate reflectance was shown in Figure 3.5. The horizontal axis is proportional to the pump-wave intensity. The behavior of the negative slope of the reflectance is unstable. Therefore, when the pump-wave intensity changes quasistatically the point of activity jumps, for example, from A to B. In addition, if the pump-wave intensity is quasistatically decreased, the activity point jumps from C to stable point D. As a result of this, the reflectance has a history and wave bistability behavior can be implemented.[19]

8.5 Making Use of Detuning Characteristics

(1) Optical-Frequency Filters

When the frequency of the probe wave is different from that of the pump wave in four-wave mixing, the reflectance shows a strong dependence on detuning[20] as explained in Section 3.2(5). This feature is being actively used in optical-frequency filters. The reflectance R is obtained from Equation (3.49). Numerical results for this equation, which represent bandpass filter characteristics, have already been shown in Figure 3.7.

Optical-frequency filters that use Na vapor as the nonlinear medium have been demonstrated with four-wave mixing.[21] Na and Ce were sealed in a cylindrical quartz cell (radius: 1.1 cm; length: 2.44 cm). The Na density was 1×10^{11} cm^{-3} due to heating. The light source was a dye laser which was circularly polarized for good separation. The optical configuration was retroreflective and the angle made by the pump and probe wave was 0.32 degrees. The pump wavelength was tuned to the $3S_{1/2}$ ($F = 2$) $\rightarrow$ $3P_{3/2}$ ($F = 3$) transition (589 nm), and the wave intensity was 200 mW/cm^2. Frequency characteristics were measured by sweeping the probe wave frequency. The full width at half maximum (FWHM) obtained here was 41 MHz and the peak reflectance was 6×10^{-5}. Not only did the pump-wave intensity decrease, but the peak reflectance was dramatically reduced even though the bandwidth did not really change.

(2) Semiconductor Laser Amplifiers

Since the end face of a semiconductor laser exhibits high reflectance, it operates according to the internal resonator format shown in Figure 6.1(g). Because there are multiple reflections within the medium the performance of effective four-wave mixing can be expected within an active medium given a collinear configuration. When the laser is operated at above the threshold value two pump waves (frequency: ω_0) propagate in mutually opposite directions within the medium. In order to promote separation from the pump wave, if we assume that an external probe wave is injected into the medium and we further assume that probe wave is shifted from the pump-wave frequency by only a small amount ($\omega_0 + \Omega$), then the probe wave having this frequency propagates in both directions within the resonator. The result of the interaction between the pump and probe waves is a phase-conjugate wave oscillating at a frequency of ($\omega_0 - \Omega$).

Incidentally, the principle of phase-conjugate wave production can be explained by resonance phenomena in absorbing media[22] and by phased diffraction gratings proportional to the pump and probe waves' beat frequency.[23]

Attempts in this regard have been made with a GaAlAs laser (wavelength: 0.83 μm; end face reflectance: 30%; CSP format).[24] Two semiconductor lasers were internally stabilized to 0.01°C and an injection current of 10 μA, and were coupled through a Faraday rotator and a half-wave plate. The light exiting the first laser was injected into the second laser's active layer; the transmitted light emanating from the second laser as well as the reflected light were detected by a PIN detector after having passed through a scanning Fabry-Perot interferometer. The waveforms were displayed on an oscilloscope. Figure 8.12 shows oscilloscope waveforms of the pump, probe and phase-conjugate waves.

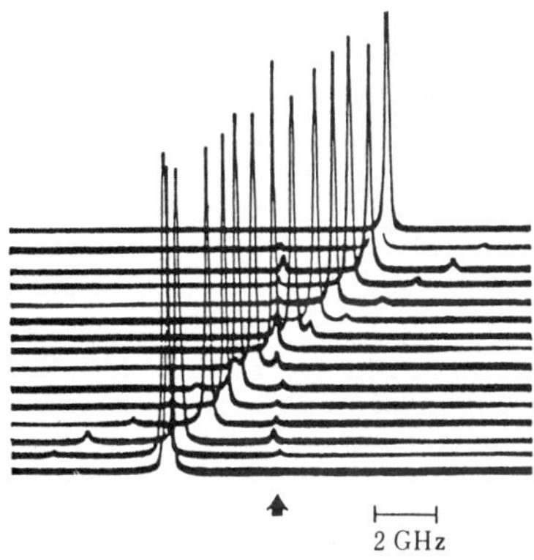

Figure 8.12 Probe, phase-conjugate and pump wave frequencies in a semiconductor laser amplifier. *(After Nakajima and Frey)*[24]

Probe-wave frequency ($\omega_0 - \Omega$) is fixed at the position of the arrow. When the pump-wave frequency ω_0 (the large signal in the diagram) is swept, a phase-conjugate wave appears at ($\omega_0 + \Omega$). The laser was GaAlAs with a wavelength of 0.83 μm.

The frequency of the probe wave was constant. The laser frequency changed at approximately −3 GHz/mA. The pump-wave frequency was swept by changing the injection current. It is clear that the phase-conjugate wave is symmetrical to the probe wave in relation to the pump-wave frequency. The energy transfer rate from the pump wave to the probe and phase-conjugate waves was 25%. An unusually high value of 5000 for phase-conjugate reflectance was obtained with several milliwatts of pump-wave power.

(3) Nonlinear Polarization

According to the explanation of resonance-type four-wave mixing given in Section 4.1(5), a diffraction grating moves at a constant velocity when there is any frequency difference between the pump and probe waves. This speed of motion competes with the atomic energy (vertical) relaxation time T_1. This type of moving diffraction grating can be used as a nonlinear polarizer for finding information within a substance. The circumstances for frequency shift include, for example

(i) when a Doppler shift is produced by translational motion within some atoms and molecules

(ii) when the external resonator's amplification bandwidth is different from that of the phase-conjugate mirror

(iii) when drift occurs during laser system operation.
When four-wave mixing has any frequency shift between the probe and pump waves it is called nearly-degenerate four-wave mixing.

As an example of nearly-degenerate four-wave mixing, consider a two-level atomic system (resonant frequency: ω_R) with a nonlinear medium having Doppler expansion. Assume that the optical system is counterpropagating and that the two pump waves (frequency: $\omega_0 = \omega_R + \Delta$) and the probe wave (frequency: $\omega_p = \omega_0 + \Omega$) are nearly collinear. Here, Δ is the shift of the pump-wave frequency from the resonant frequency and Ω is the shift of the probe wave from the resonant frequency. If Δ and ω_0 are fixed and if ω_p is swept, the conditions for resonance are: (i) $\Omega = 0$ when the forward pump and probe waves form a diffraction grating and the back pump wave is scattered, and (ii) $\Omega = 2\Delta$ when each light ray receives a Doppler shift, when the forward pump wave has atomic resonance, when the frequencies of the back pump and probe waves agree, and when a diffraction grating is formed.

Figure 8.13 is a measured example of (Δ, Ω) shift dependency when Na vapor is used as the nonlinear medium.[25]

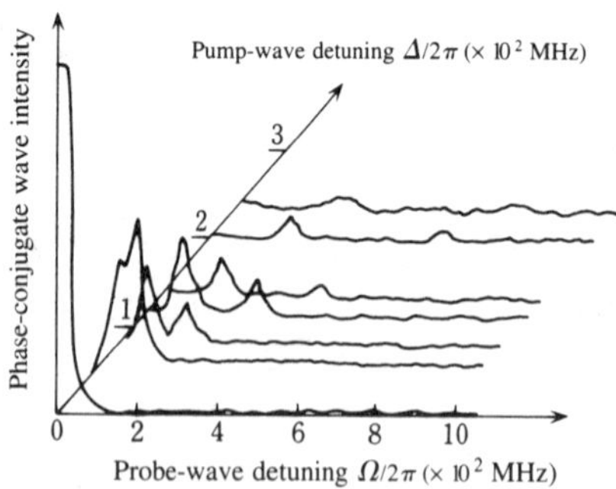

Figure 8.13 Relationship between phase-conjugate wave intensity and detuning in Na vapor. *(After Steel and Lind)*[25]

Wavelength: in the neighborhood of 589 nm; pump-wave frequency $\omega_0 = \omega_R + \Delta$; probe-wave frequency: $\omega_p = \omega_0 + \Omega$; ω_R = atomic resonance frequency.

The light source was a double dye laser (linewidth: of the order of 1 MHz) tuned to the neighborhood of the D$_2$ line of the Na atom (wavelength: 589 nm). The spectrum was observed by a scanning Fabry-Perot interferometer. When $\Delta = 0$

and Ω is in the vicinity of 0, one strong resonance line appears. Two resonance peaks emerge at $\Omega = 0$ and $\Omega = 2\Delta$ when Δ is increased. As explained in Section 4.1(5), the energy (vertical) relaxation frequency can be found from the central peak width. A frequency of 25 MHz was obtained from the data in the figure; this agreed well with the expected value of 20 MHz obtained from a separate measurement.

The ac Stark effect was also observed by the same type of measurement.[25]

8.6 Use of Temporal Parity (Group Delay Equalizer)

One of the characteristics of phase-conjugate light is temporal parity. An example of the use of a group delay equalizer for this feature is shown in Figure 8.14.

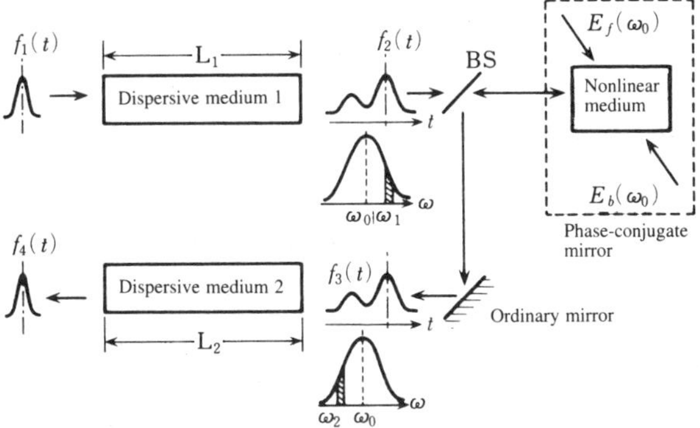

Figure 8.14 Application of phase-conjugate waves to equalize group delay.

Injected wave pulse $f_1(t)$ becomes broadened output pulse $f_2(t)$ by dispersion in medium 1. If this is injected into a phase-conjugate mirror, the frequency spectrum of output pulse $f_3(t)$ is opposite that of $f_2(t)$. If $f_3(t)$ propagates in medium 2 having the same characteristics as medium 1, the original narrow wave pulse $f_4(t)$ is obtained. E_f, E_b: pump waves; $\omega_1 - \omega_0 = \omega_0 - \omega_2$.

If an injected light pulse $f_1(t)$ propagates in a dispersive medium, the pulse width is broadened due to group velocity dispersion. If this pulse is injected into a phase-conjugate mirror, then the frequency component of the outgoing phase-conjugate light is reversed. If this phase-conjugate light then propagates through a dispersive medium, the contribution of group velocity dispersion is cancelled if certain conditions are satisfied, and the original narrow light pulse is reconstituted.[27] We next show an equation that can be used to explain the principle of operation.

The following equation represents the incident light pulse $f_1(t)$:

$$f_1(t) = g(t)\exp(i\omega_0 t) \tag{8.21}$$

where $g(t)$ is the envelope, and ω_0 is the frequency of the optical carrier wave. If the Fourier transform of the envelope is $G(\Omega)$, then

$$g(t) = \int_{-\infty}^{\infty} G(\Omega) \exp(i\Omega t) \, d\Omega \tag{8.22}$$

from which the injected pulse can be written as follows:

$$f_1(t) = \int_{-\infty}^{\infty} G(\Omega) \exp[i(\omega_0 + \Omega) t] \, d\Omega \tag{8.23}$$

In practice, since the injected pulse's spectral width is considerably smaller than the frequency of the carrier wave, it can be concluded that $\Omega \ll \omega_0$. The pulse's waveform at the point of exit is given by

$$f_2(t) = \int_{-\infty}^{\infty} G(\Omega) \exp\{i[(\omega_0 + \Omega) t - \beta_1(\omega_0 + \Omega) L_1]\} \, d\Omega \tag{8.24}$$

for a pulse of length L_1 propagating through a dispersive medium having propagation constant $\beta_1(\omega)$. The Taylor series expansion for the propagation constant β_1 in the neighborhood of ω_0 becomes

$$\beta_1(\omega_0 + \Omega) = \beta_1(\omega_0) + \beta_1^{(1)}\Omega + (1/2)\beta_1^{(2)}\Omega^2 + \cdots \tag{8.25}$$

where $\beta_1^{(n)} \equiv \partial^n \beta_1 / \partial \omega^n$ is the nth derivative. If we consider Equation (8.24) up to the second derivative, then the outgoing pulse can be written as follows:

$$f_2(t) = \exp[i(\omega_0 t - \beta_{10} L_1)] \int_{-\infty}^{\infty} G(\Omega)$$
$$\cdot \exp\{i[\Omega(t - \beta_1^{(1)} L_1) - (1/2)\beta_1^{(2)}\Omega^2 L_1]\} \, d\Omega \tag{8.26}$$

where $\beta_{10} \equiv \beta_1(\omega_0)$, $\beta_1^{(1)}$ is the reciprocal of the group velocity, and $\beta_1^{(2)}$ is the group velocity dispersion. The optical pulse receives a delay due to the group velocity, and the pulse width broadens due to group velocity dispersion.

When an outgoing pulse f_2 from a dispersive medium is injected into a phase-conjugate mirror, this outgoing pulse is represented by f_3. The phase-conjugate mirror implements degenerate four-wave mixing with a pump-wave frequency of ω_0. If this is done, from the results on detuning in Section 3.2(5), the $(\omega_0 + \Omega)$ probe-wave frequency component converts to a phase-conjugate wave having frequency $(\omega_0 - \Omega)$. Now, if the phase-conjugate mirror's complex reflectance at frequency $(\omega_0 + \Omega)$ is represented by $r(\Omega) \exp[i\phi(\Omega)]$, then f_3 can be expressed as follows:

$$f_3(t) = \exp[i(\omega_0 t + \beta_{10} L_1)] \int_{-\infty}^{\infty} r(\Omega) G^*(\Omega)$$
$$\cdot \exp\{i[-\Omega(t - \beta_1^{(1)} L_1) + (1/2)\beta_1^{(2)}\Omega^2 L_1 + \phi(\Omega)]\} \, d\Omega \tag{8.27}$$

If this wave propagates in a second dispersive medium (propagation constant: β_2; length: L_2), we obtain the following Taylor series expansion for the outgoing waveform where the propagation constant is β_2 in the neighborhood of ω_0:

$$f_4(t) = \exp[i(\omega_0 t + \beta_{10}L_1 - \beta_{20}L_2)] \int_{-\infty}^{\infty} r(\Omega)\, G^*(\Omega)$$
$$\cdot \exp\{i[-\Omega(t - \beta_1^{(1)}L_1 - \beta_2^{(1)}L_2)$$
$$+ (1/2)(\beta_1^{(2)}L_1 - \beta_2^{(2)}L_2)\,\Omega^2 + \phi(\Omega)]\}\,d\Omega \tag{8.28}$$

where $\beta_{20} \equiv \beta_2(\omega_0)$. When looking at Equation (8.28) we see that although a group velocity term is appended, the group velocity dispersion term is cancelled out if

$$(\partial^2 \beta_1 / \partial\omega^2)\, L_1 = (\partial^2 \beta_2 / \partial\omega^2)\, L_2 \tag{8.29}$$

is satisfied. That is, the broadening of the pulse width is compensated for if the pulse passes twice through a medium that satisfies Equation (8.29) and has the same group velocity dispersion values. Attention should be given to the fact that f_2 and f_3 have an inverse phase relationship here. The waveforms, however, are not inverses.

When numerical values are calculated, broadening of the pulse can be cancelled when a phase-conjugate mirror having a phase characteristic of $\phi(\Omega)$ is used with pulse widths up to 200 ps because there is almost no deviation from a straight line in the area where $\Omega < 4$ GHz for a phase-conjugate mirror.[27]

8.7 Application of Time-Domain Multiplicative Interactions

(1) Optical Pulse Compression in Forward Four-Wave Mixing

We will first explain the principle of optical pulse compression. As shown in Figure 8.15, we assume two optically intense pulsed pump waves E_1 and E_2 coming from the same side of the nonlinear medium, and injected into it at a small angle from each other.

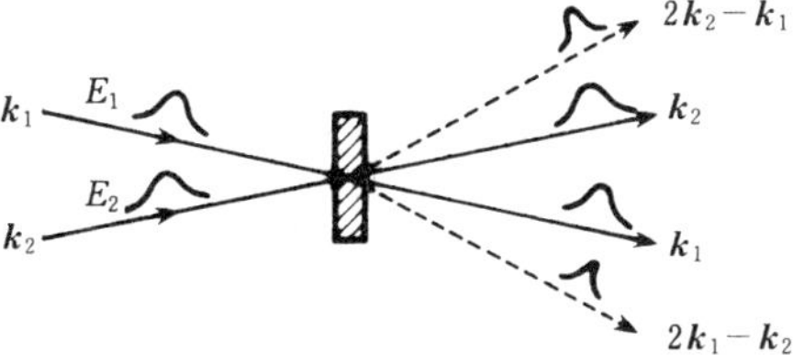

Figure 8.15 Principle of optical pulse compression in forward four-wave mixing. *(After Fujimoto and Ippen)*[29]

We take their respective electric fields to be $A_j(t)$, and their wave vectors to be $\boldsymbol{k}_j$, and assume the two pulse time intervals to be ΔT (relative input wave pulse FWHM). The relationship between the input pulse width t_w, the medium's energy (vertical) relaxation time T_1, and the phase (horizontal) relaxation time T_2 is assumed to become $T_2 < t_w < T_1$. Given this, the substance excitation wave is proportional to the product $A_1 A_2{}^*$ of the electric fields and, because of this, A_1 self-diffracts and scatters in the $2\boldsymbol{k}_1 - \boldsymbol{k}_2$ direction.[28] The polarization created by the interaction of the electric field and the nonlinear medium can be expressed by

$$P(t, \Delta T) \propto A_1(t) \int_{-\infty}^{\infty} A_1(t') A_2{}^*(t' + \Delta T)\, dt'$$

(8.30)

Therefore, if the pulse time interval ΔT is adjusted, a pulse width narrower than that of the input pulse can be obtained. Dispersion of the light pulse in the $2\boldsymbol{k}_2 - \boldsymbol{k}_1$ direction can be explained in a similar way.

We explain next an experiment in which the above-mentioned principle was used.[29] The light source was a colliding pulse mode, mode-locked ring dye laser (rhodamine 6G). The exciting wave was a continuous argon ion laser, passive mode-synchronized with absorptive DODCI. The laser output was an approximately 68-fs-wide pulse at a wavelength of 620 nm, having a pulse energy of 0.15 nJ. A light pulse with a width of 65 fs and 50 μJ of energy was obtained through three-stage dye amplification (Nd:YAG laser excitation). After this beam was branched by a beam splitter, an optical delay was performed mechanically. The two beams were injected into a nonlinear cell from the same side, with the angle between the beams set at 1.5 degrees. Malachite green dissolved in methanol was placed in the cell whose thickness was 0.1 mm. When the relative pulse interval was $\Delta T = -1.8$, the pulse width was 37 fs, with a maximum compression rate of 0.54. The maximum value obtained for phase-conjugate reflectance was approximately 3 percent.

(2) Optical Pulse Compression by Stimulated Brillouin Scattering

Optical pulse compression by stimulated Brillouin scattering makes use of a tapered waveguide. The principle of optical pulse compression will be explained by using Figure 8.16.[30] The length of the tapered waveguide is taken such that the optical pulse covers the whole taper; the thickness is such that the stimulated Brillouin scattering threshold wave intensity occurs at the midpoint of the waveguide. If light is injected from the thick end of the taper, the optical intensity increases as the wave propagates. When the optical intensity exceeds the threshold for stimulated Brillouin scattering, back-scattered waves begin to be produced and the gain increases. The phonon intensity increases in concert with the growth of the back-scattered waves, and the phonons propagate at the speed

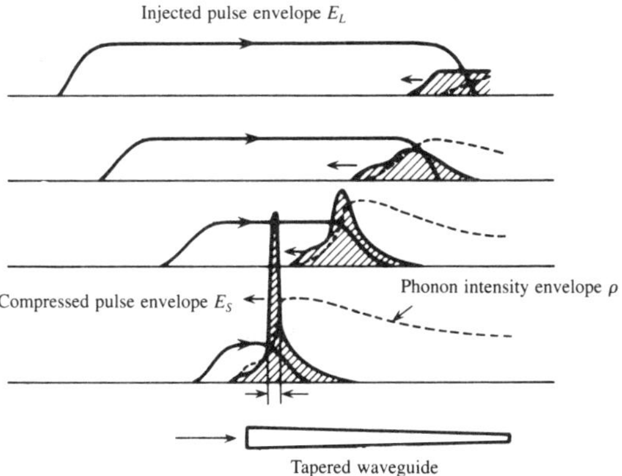

Figure 8.16 Explanatory diagram of the principle of optical pulse compression in stimulated Brillouin scattering using a tapered waveguide. *(After Hon)*[30]

of light within the medium. Consequently, phase-conjugate wave reflectance can be seen to increase in regard to back-scattered pulses propagating in the direction of the thickness of the taper. The result is that the light pulses are compressed in the time domain.

The experiment was performed as follows. The optical source was a vertical single-mode Nd:YAG laser (wavelength: 1.06 μm; pulse width: 20 ns; pulse energy: 200 mJ). The nonlinear medium was methane gas (CH_4) sealed in a glass tube at a pressure of 130 atmospheres. The glass tube was 1.3 m long, and gradually tapered from an inside diameter of 4 mm to one of 2 mm. The optical pulse width was compressed from 20 ns to 2 ns.

8.8 Applications of Quantum Correlation (Producing Optically-Squeezed States)

If light is treated as a quantum phenomenon, then the electric field complex amplitudes A and A^* correspond to the photon annihilation and creation operators $\hat{a}$ and $\hat{a}^\dagger$, respectively. These quantum fluctuations are not independent, but have a certain constant correlation. If we take advantage of this characteristic, measurements can be made down to very small signals. We will take up an example of the optically-squeezed state in this section.

A quantum opto-electric domain can be expressed as follows:[31]

$$E(t) = \xi[\hat{a}\exp(-i\omega t) + \hat{a}^{\dagger}\exp(i\omega t)]$$
$$= 2\xi[\hat{X}_1\cos(\omega t) + \hat{X}_2\sin(\omega t)] \tag{8.31}$$

where $\hat{a}$ and $\hat{a}^{\dagger}$ are the annihilation and creation operators, respectively, $\hat{X}_1 \equiv (\hat{a} + \hat{a}^{\dagger})/2$ and $\hat{X}_2 \equiv (\hat{a} - \hat{a}^{\dagger})/2i$ are the Hermitian operators of the orthogonal phase components, and ξ is a constant of proportionality. These operators satisfy the following transformation functions:

$$[\hat{a}, \hat{a}^{\dagger}] = 1, \quad [\hat{X}_1, \hat{X}_2] = i/2 \tag{8.32}$$

We obtain the following equation from the transformation functions for the uncertainty function for $\hat{X}_1$ and $\hat{X}_2$:

$$\langle(\Delta\hat{X}_1)^2\rangle\langle(\Delta\hat{X}_2)^2\rangle \geq 1/16 \tag{8.33}$$

where

$$\langle(\Delta\hat{X}_m)^2\rangle = \langle(\hat{X}_m - \langle\hat{X}_m\rangle)^2\rangle \quad (m = 1, 2) \tag{8.34}$$

is the dispersion (quantum fluctuation), and < > represents quantum averaging. Equation (8.33) says that the product of the uncertainties of the two orthogonal phase components is always larger than some constant value. When the equality holds (minimum uncertainty), this formula describes the coherent state. The annihilation operators in the coherent state are such that the uncertainties of $\hat{X}_1$ and $\hat{X}_2$ are equal:

$$\langle(\Delta\hat{X}_1)^2\rangle = \langle(\Delta\hat{X}_2)^2\rangle = 1/4 \tag{8.35}$$

Incidentally, Relation (8.33) can be an equality even if the magnitudes of the the two terms on the left-hand side are not equal. The uncertainties of the two orthogonal phase components can vary while the uncertainty product remains equal to the minimum. This situation is called a squeezed state (the two-photon coherent state is a special case).[32,33] Here,

$$\langle(\Delta\hat{X}_m)^2\rangle < 1/4 \quad (m = 1 \text{ and } 2) \tag{8.36}$$

This means that smaller measurement uncertainties can be achieved by observing one orthogonal phase component of a squeezed state than those available with coherent-state light. The squeezed state can be represented by

$$\hat{b} = \mu\hat{a} + \nu\hat{a}^{\dagger} \tag{8.37}$$

(single-mode case), where $|\mu|^2 - |\nu|^2 = 1$. A diagram of the probability density distributions for the orthogonal phase components in coherent and squeezed states is shown in Figure 8.17.

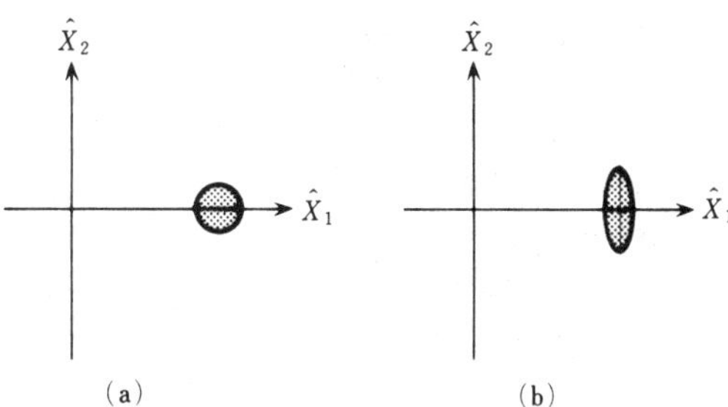

Figure 8.17 Probability density distribution of quantum-optical orthogonal phase components.
(a) Coherent state; (b) squeezed state.

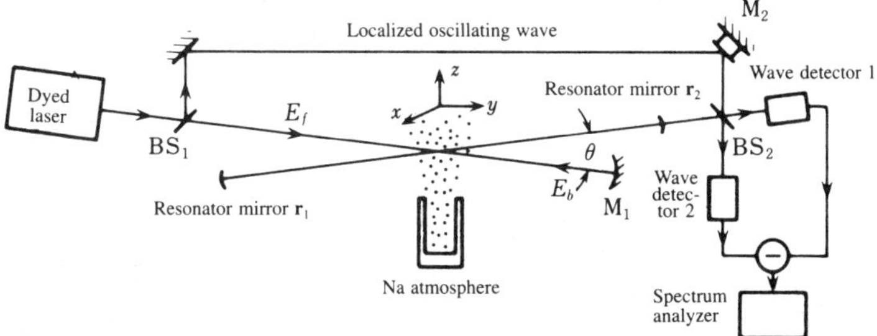

Figure 8.18 Experimental setup for producing optically-squeezed states by four-wave mixing. *(After Slusher, Hollberg, Yurke, Mertz, and Valley)*[35]

E_f, E_b: forward and back pump waves;
M$_2$: piezo-driven mirror.

In line with Equation (8.37), $\hat{a}^{\dagger}$ is the phase-conjugate term corresponding to $\hat{a}$, and ties together the optically-squeezed state and the phase-conjugate wave. It has been shown theoretically that an optically-squeezed state can be produced by using degenerate four-wave mixing.[34] See Appendix D(1).

Figure 8.18 shows the experimental setup for producing optical squeezing.[35] The nonlinear medium was Na vapor with a density of 10^{11} cm^{-3} and an interaction length of 1 cm; the D$_2$ resonance beam's linear absorption coefficient was 5 cm^{-1}. The exciting source was an argon ion laser (wavelength: 589 nm; beam power: 0.5 W), the pump wave was tuned to a frequency $1-2$ GHz lower than the D$_2$ resonance line; the absorption coefficient was 10^{-3} cm^{-1} at the pump frequency. The optical setup was retroreflective, with an 0.86-degree angle between the pump wave and the resonator (formed by r_1 and r_2). Vacuum fluctuations within the resonator gave rise to the probe wave and a wave phase-conjugate to it was generated. A squeezed state was produced by the phase-conjugate wave that was reflected by r_1 and by the probe wave passing directly through the Na. The reflected wave from beam splitter BS_1 was used as a localized oscillating wave and was homodyne detected. Two modes — above and below the localized oscillating frequency — were produced in this way by the localized oscillating wave and the optically squeezed state, with a quantum relationship between the modes. Accordingly, if the phase of the localized oscillating wave was changed by a piezo-driven mirror M_2, a specific phase could be attached to one of the modes since there is a quantum relationship. Less than 0.3 dB of shot noise was observed.

Forward four-wave mixing was used with quartz optical fiber[36] and Na vapor[37] as the nonlinear media in experiments to produce optically-squeezed states.

Conclusion

Given a 1972 starting point for phase-conjugate optics, it appears that all ingredients of basic principles, assorted experimental examples and various proposals for applications have emerged in outline over approximately twenty years. However, the significance of this lies not only in the implementation of an optical discipline whose concept corresponds to complex mathematical conjugation, but in the possibilities for future research that will enable the remarkable properties of phase-conjugate light to be truly useful to humankind.

Phase-conjugate optics is not the sole area of nonlinear optics; other optical phenomena such as holography, photon echoes, optical bistability, optical chaos, and stimulated optical scattering are intimately related. In addition, in recent years there have been striking developments with such highly nonlinear media as nonlinear organic materials and quantum-well structures. Phase-conjugate optics is making best use of these points of contact, with reciprocal advances in understanding and technology taking place in related disciplines. In this regard, phase-conjugate optics takes advantage of unique features of phase-conjugate light media, in which steady progress has been made. Phase-conjugate optics is simply a branch of nonlinear optics, and comprises an independent research area. More progress is anticipated.

To conclude this book, we next enumerate major research issues for the future.

(1) Direct measurement of electric field phase: Although phase-conjugate wave phase correction is currently being used to verify the production of phase-conjugate light, it is an indirect method. Direct comparison of the probe field and the phase-conjugate field would be ideal. This is also essential to the experimental measurement of the fidelity of phase-conjugate waves as described next.

(2) Clarifying the conditions for improving the level of fidelity of phase-conjugate light: The phase-conjugate wave propagation functions described in Section 4.4 are also in this category. Concentrated research efforts will also be made possible by determining the relationships between various light-source and medium parameters, as well as that with the fidelity level of phase-conjugate light. For example, there is the possibility of planning for the improvement of temporal stability, one of the current application issues. Also, since light-source coherence and materials can be classified by application type, futile investments can be avoided while application flexibility is increased.

(3) Making $\chi^{(3)}$ large and developing fast-response-time media: Generally, response time is slow when $\chi^{(3)}$ is large in currently available media, producing a trade-off between the value of $\chi^{(3)}$ and the response time. However, it is also clear that there is a dimensional relationship for materials vis-à-vis large values of $\chi^{(3)}$ implemented through macromolecular single crystals and multi-quantum-well semiconductors. This, we believe, will not only involve simply investigating materials as such, but the design of materials for stated objectives will become necessary in the future.

(4) Discovering applications in which phase-conjugate light has an essential role: Although many applications for phase-conjugate light have been proposed, few have served practical applications up to now. In addition to alternatives to conventional technology, improved technology must be provided. Because of this, phase-conjugate wave reflectance, response speed, etc., must improve and improvement in reliability through such means as increasing temporal stability must be researched.

Appendix

A. Relationship Between the SI and cgs esu Systems of Units

Although the area of electromagnetics has historically developed through the cgs unit system, the SI (MKSA; international) system of units was introduced as a result of the reformulation of theoretical systems. The SI system has recently come into use in academic circles. Because of this, the theory presented in this book, too, is described in SI units. However, when it comes to actual discussions, the cgs system is used overwhelmingly for experiments, while the SI system is mixed with the cgs system in discussions of theoretical relationships. Neophytes can frequently become ensnared in a confusion of unit systems. The relationships between both systems of units are shown in this Appendix, the objective of which is to accommodate the reader.

(1) Wave Equations, Nonlinear Polarization, and Expressions for Wave Intensity

The wave equations for both systems of units are given by the following in the case of a non-absorptive medium:

$$\nabla^2 E - \frac{1}{c^2} \cdot \frac{\partial^2 (\varepsilon E)}{\partial t^2} = \mu_0 \frac{\partial^2 P_{NL}}{\partial t^2} \qquad : \text{SI} \tag{A.1}$$

$$\nabla^2 E - \frac{1}{c^2} \cdot \frac{\partial^2 (\varepsilon E)}{\partial t^2} = \frac{4\pi}{c^2} \cdot \frac{\partial^2 P_{NL}}{\partial t^2} \qquad : \text{cgs} \tag{A.2}$$

where E is the electric field, P_{NL} is the nonlinear polarization, ε is the electric permittivity, $c = 1/(\varepsilon_0 \mu_0)^{1/2}$ is the speed of light in vacuo, ε_0 is the permittivity of free space, and μ_0 is the in-vacuo magnetic permeability. The following two equations give the nonlinear polarization:

$$P_{NL} = \overline{\chi}^{(1)} E + 2D_2 \overline{\chi}^{(2)} E^2 + 4D_3 \overline{\chi}^{(3)} E^3 + \cdots \qquad : \text{SI, cgs} \tag{A.3}$$

$$P_{NL} = \varepsilon_0 (\chi^{(1)} E + 2D_2 \chi^{(2)} E^2 + 4D_3 \chi^{(3)} E^3 + \cdots) \qquad : \text{SI} \tag{A.4}$$

where the D_n are the degeneracy factors related to frequency and show the effect of different combinations of E on the same P_{NL} component as a function of permutations of E.[1] For example, $D_3 = 3$ in degenerate four-wave mixing. $\chi^{(n)}$ is the nth order nonlinear susceptibility, written with a bar (—) when ε_0 does not appear in the equation. Although Equation (A.3) is often used for both unit systems, Equation (A.4) is employed only for SI units. In addition, wave intensity I is described by the following equations:

$$I = nc\varepsilon_0 E^2/2 \quad : \text{SI} \tag{A.5}$$

$$I = ncE^2/8\pi \quad : \text{cgs} \tag{A.6}$$

where n is the refractive index of the medium.

(2) Conversions for Third-Order Nonlinearity

We show here conversion expressions for the two types of third-order nonlinearity. The relationship between the SI and cgs esu systems for the nonlinearity in Equation (A.3) is given by

$$\overline{\chi}^{(3)} \, [\text{Fm}/\text{V}^2] = [1/(3^4 \times 10^{17})] \, \overline{\chi}^{(3)} \quad [\text{esu}] \tag{A.7}$$

In addition, the relationship between nonlinearity in Equations (A.3) and (A.4) is given by

$$\chi^{(3)} \, [\text{m}^2/\text{V}^2] = [4\pi/(3^2 \times 10^8)] \, \overline{\chi}^{(3)} \quad [\text{esu}] \tag{A.8}$$

(3) Relationship Between Third-Order Nonlinearity and Nonlinear Refractive Index

Although the refractive index n of a medium can be described by the linear refractive index n_0 when the wave intensity is weak, a change in the refractive index proportional to the intensity I is produced when a strong source like a laser is employed. The optical Kerr effect is an example of this. This coefficient of proportionality for the change in the refractive index is called the nonlinear refractive index n_2, and is defined by the following two formulas:

$$n = n_0 + n_2 I \tag{A.9}$$

$$n = n_0 + n_2 \langle E^2 \rangle = n_0 + n_2 A^2/2 \tag{A.10}$$

where $< >$ denotes time averaging and A is the complex electric field amplitude. The nonlinear refractive index here is related to the real part of the third-order nonlinearity, $\chi_R^{(3)}$. Since $D_3 \chi_R^{(3)} E^2$ is equivalent to the change in relative permittivity $\delta\varepsilon$ when Definitions (A.4) and (A.9) are used, $\delta\varepsilon = 2n_0\delta n = 2n_0 \cdot n_2 I$ (δn: change in refractive index) is used with SI units and we obtain

$$n_2 \, [\text{m}^2/\text{W}] = [3/(c\varepsilon_0 n_0^2)] \chi_R^{(3)} (\omega \,; \omega, \omega, -\omega) \quad [\text{m}^2/\text{V}^2] \tag{A.11}$$

In addition,

$$n_2 \ [\text{m}^2/\text{V}^2] = (3/n_0)\, \chi_R^{(3)}(\omega\,;\,\omega,\,\omega,\,-\omega) \quad [\text{m}^2/\text{V}^2] \tag{A.12}$$

is obtained when Definitions (A.4) and (A.10) are used in the same way. When Equations (A.3) and (A.10) are used with cgs esu units we obtain

$$n_2 \ [\text{esu}] = (12\pi/n_0)\, \overline{\chi}_R^{(3)}(\omega\,;\,\omega,\,\omega,\,-\omega) \quad [\text{esu}] \tag{A.13}$$

since $4\pi D_3 \overline{\chi}^{(3)} E^2$ corresponds to the change in relative permittivity. Because the linear refractive index appears in the conversion formulas, n_2 and $\chi^{(3)}$ cannot be replaced by a constant multiple. However, since the linear refractive index n_0 may be at most 3, n_2 and $\chi^{(3)}$ in the cgs system can be estimated to have a difference of approximately one order of magnitude. This assumes that degenerate four-wave mixing is used in Equations (A.11) to (A.13) and that $D_3 = 3$.

The following equation can be used to convert units for the nonlinear refractive index n_2:

$$\begin{aligned}
n_2 \ [\text{m}^2/\text{V}^2] &= [1/(3^2 \times 10^8)]\, n_2 \quad [\text{esu}] \\
&= (n_0/120\pi)\, n_2 \quad [\text{m}^2/\text{W}] \tag{A.14}
\end{aligned}$$

Table A lists the conversion factors discussed above. The SI values for the nonlinear refractive index n_2 are not strictly SI units, but employ practical units [cm²/W]. As an example of how the table is used, if we assume that

Table A Conversion factors for third-order nonlinearity $\chi^{(3)}$ and nonlinear refractive index n_2

	$\chi^{(3)}$ [m²/V²]	$\overline{\chi}^{(3)}$ [Fm/V²]	$\overline{\chi}^{(3)}$ [esu]	$n_2 \cdot n_0$ [m²/V²]	$n_2 \cdot n_0^2$ [cm²/W]	$n_2 \cdot n_0$ [esu]
$\chi^{(3)}$ [m²/V²]	1	$\dfrac{1}{4\pi \times 3^2 \times 10^9}$	$\dfrac{3^2 \times 10^8}{4\pi}$	3	$36\pi \times 10^5$	$3^3 \times 10^8$
$\overline{\chi}^{(3)}$ [Fm/V²]	$4\pi \times 3^2 \times 10^9$	1	$3^4 \times 10^{17}$	$4\pi \times 3^3 \times 10^9$	$16\pi^2 \times 3^4 \times 10^{14}$	$12\pi \times 3^4 \times 10^{17}$
$\overline{\chi}^{(3)}$ [esu]	$\dfrac{4\pi}{3^2 \times 10^8}$	$\dfrac{1}{3^4 \times 10^{17}}$	1	$\dfrac{4\pi}{3 \times 10^8}$	$\dfrac{16\pi^2}{10^3}$	12π
$n_2 \cdot n_0$ [m²/V²]	$\dfrac{1}{3}$	$\dfrac{1}{4\pi \times 3^3 \times 10^9}$	$\dfrac{3 \times 10^8}{4\pi}$	1	$12\pi \times 10^5$	$3^2 \times 10^8$
$n^2 \cdot n_0^2$ [cm²/W]	$\dfrac{1}{36\pi \times 10^5}$	$\dfrac{1}{16\pi^2 \times 3^4 \times 10^{14}}$	$\dfrac{10^3}{16\pi^2}$	$\dfrac{1}{12\pi \times 10^5}$	1	$\dfrac{3 \times 10^3}{4\pi}$
$n_2 \cdot n_0$ [esu]	$\dfrac{1}{3^3 \times 10^8}$	$\dfrac{1}{12\pi \times 3^4 \times 10^{17}}$	$\dfrac{1}{12\pi}$	$\dfrac{1}{3^2 \times 10^8}$	$\dfrac{4\pi}{3 \times 10^3}$	1

When the relationship between $\chi^{(3)}$ and n_2 is to be found, the level of degeneracy in nonlinear polarization is assumed to be for degenerate four-wave mixing and we further assume that $D_3 = 3$. In addition, n_0 is the linear refractive index.

$n_0 = 1.63$ when the nonlinear refractive index $n_2 = 1.2 \times 10^{-11}$ esu for CS_2, then $\bar{\chi}^{(3)} = n_2 n_0 / 12\pi = 5.2 \times 10^{-13}$ esu.

B. Third-Order Nonlinearity and Symmetry of Electro-Optic Tensors and Crystals

Third-order nonlinearity $\chi^{(3)}$ is often made use of in four-wave mixing. It exists as a rank-4 tensor where the nonzero components are determined by the symmetry of the crystal. Third-order nonlinear polarization is written as

$$P_m(\omega_4) = (1/2)\,\varepsilon_0 \sum_{n,o,p} D_3\, \chi^{(3)}_{mnop} A_n(\omega_1)\, A_o(\omega_2)\, A_p(\omega_3) + \text{c.c.} \tag{B.1}$$

where the A_j are the complex electric field amplitudes; m, n, o, and p represent any of the orthogonal coordinates x, y, and z within the crystal; and D_3 is the degeneracy. The relationship between the nonzero components of $\chi^{(3)}$ determined by crystal symmetry and the several components for highly symmetric crystal structures often used in experiments is presented in Table B.1.[2,3]

The Pockels effect is related to crystal symmetry in the photorefractive effect. As shown in Equation (3.62), the relative permittivity can be expressed by

Table B.1 Nonzero components of $\chi^{(3)}{}_{mnop}$ $(\omega_4; \omega_1, \omega_2, \omega_3)$ for any crystal system[2,3]

Crystal system		Independent nonzero components
Isometric crystal system		$xxxx = yyyy = zzzz = xxyy + xyxy + xyyx$ $yyzz = zzyy = zzxx = xxzz = xxyy = yyxx$ $yzyz = zyzy = zxzx = xzxz = xyxy = yxyx$ $yzzy = zyyz = zxxz = xzzx = xyyx = yxxy$
Cubic crystal structure	23, m 3	$xxxx = yyyy = zzzz$ $yyzz = zzxx = xxyy,\quad zzyy = yyxx = xxzz$ $yzyz = zxzx = xyxy,\quad zyzy = xzxz = yxyx$ $yzzy = zxxz = xyyx,\quad zyyz = xzzx = yxxy$
	432, $\bar{4}3$ m, m 3 m	$xxxx = yyyy = zzzz$ $yyzz = zzyy = zzxx = xxzz = xxyy = yyxx$ $yzyz = zyzy = zxzx = xzxz = xyxy = yxyx$ $yzzy = zyyz = zxxz = xzzx = xyyx = yxxy$
Orthorhombic crystal structure 422, 4 mm, 4/mmm, $\bar{4}2$ m		$xxxx = yyyy,\quad zzzz$ $yyzz = zzyy,\quad zzxx = xxzz,\quad xxyy = yyxx$ $yzyz = zyzy,\quad zxzx = xzxz,\quad xyxy = yxyx$ $yzzy = zyyz,\quad zxxz = xzzx,\quad xyyx = yxxy$
Hexagonal crystal structure 6 mm, 622, 6/mmm, $\bar{6}$ m 2		$zzzz,\quad xxxx = yyyy = xxyy + xyyx + xyxy$ $xxyy = yyxx,\quad xyyx = yxxy,\quad xyxy = yxyx$ $yyzz = xxzz,\quad zzyy = zzxx,\quad zyyz = zxxz$ $yzzy = xzzx,\quad yzyz = xzxz,\quad zyzy = zxzx$

$$\varepsilon_\omega = \varepsilon_n - \varepsilon_n (\vec{R} E_{st})\, \varepsilon_n \tag{B.2}$$

where ε_ω is the permittivity tensor of the wave domain, ε_n is the permittivity tensor when there is no electric field, E_{st} is a very slowly-varying electric field compared to the wave frequency, and $\vec{R}$ is the rank-3 electro-optic tensor. The following can be written now if we assume that $1 = x$, $2 = y$, and $3 = z$:

$$(\vec{R} E_{st}) = \begin{pmatrix} \varepsilon_1 & \varepsilon_6 & \varepsilon_5 \\ \varepsilon_6 & \varepsilon_2 & \varepsilon_4 \\ \varepsilon_5 & \varepsilon_4 & \varepsilon_3 \end{pmatrix}, \quad \begin{pmatrix} \varepsilon_1 \\ \varepsilon_2 \\ \varepsilon_3 \\ \varepsilon_4 \\ \varepsilon_5 \\ \varepsilon_6 \end{pmatrix} = \begin{pmatrix} r_{11} & r_{12} & r_{13} \\ r_{21} & r_{22} & r_{23} \\ r_{31} & r_{32} & r_{33} \\ r_{41} & r_{42} & r_{43} \\ r_{51} & r_{52} & r_{53} \\ r_{61} & r_{62} & r_{63} \end{pmatrix} \begin{pmatrix} E_1 \\ E_2 \\ E_3 \end{pmatrix} \tag{B.3}$$

Electro-optic tensors for crystal structures often used in experiments are shown in Table B.2.[4]

Table B.2 Electro-optic tensor $\vec{R}$ for any crystal group[4]

Crystal structure	Electro-optic tensor	
	$\bar{4}3\,\mathrm{m},\ 23$	$432,\ \mathrm{m}\,3,\ \mathrm{m}\,3\,\mathrm{m}$
Cubic crystal structure		
	$4\,\mathrm{mm}$	$\bar{4}2\,\mathrm{m}\,(2 \parallel x_1)$
Orthorhombic crystal structure		
	$6\,\mathrm{mm}$	622
Hexagonal crystal structure		
	$3\,\mathrm{m}\,(\mathrm{m}\perp x_1,\ \text{standard orthography}$	$3\,\mathrm{m}\,(\mathrm{m}\perp x_2)$
Trigonal crystal structure		

Note: $\cdot$: zero component; $\bullet$: nonzero component; $\bullet\!\!-\!\!\bullet$: equal nonzero values; $\bullet\!\!-\!\!\circ$: equal absolute values but opposite in sign; $\circledcirc$: nonzero component equal to -2 times the linked component.

C. Phase-Conjugate Wave Analysis Using Lagrangians

We explain here phase-conjugate wave analysis techniques that employ Lagrangians often used in mechanics. We will deal with the case of Section 3.2(4) where pump-wave depletion was analyzed in terms of the optical Kerr effect.

We write the electric fields and nonlinear polarization as follows:

$$\boldsymbol{E} = (1/2)\{\sum A_m [i(\omega t - \boldsymbol{k} \cdot \boldsymbol{r})]\boldsymbol{e}_m + \text{c.c.}\}$$
$$\boldsymbol{P}_{NL} = (1/2)\,\varepsilon_0\{\sum P_m \exp[i(\omega t - \boldsymbol{k} \cdot \boldsymbol{r})]\boldsymbol{e}_m + \text{c.c.}\} \quad (m = x, y \text{ or } z)$$

$$(\text{C.1})$$

where the A_m are the complex amplitude components of the electric fields, the P_m are the complex amplitude components of the polarizations, $\boldsymbol{k}$ is the wave vector, $\boldsymbol{r}$ is the position vector, ε_0 is the permittivity of free space, and $\mathbf{e}_m$ is the unit vector. We assume the direction of propagation to be the z axis. Substituting the above equations into Wave Equations (3.16) and using an adiabatic approximation, we obtain the following equation for each orthogonal component:

$$-i\frac{dA_m}{dz} + \frac{\omega}{2cn} P_m = 0$$

$$(\text{C.2})$$

where n is the refractive index and c is the speed of light. Equation (C.2) resembles the linkage equations of mechanics, and we pick out the Lagrangian that satisfies this equation. The first term of Equation (C.2) is the dissipative term[5] for which the corresponding Lagrangian is

$$\mathscr{L}_1 = -(1/2)[(-idA/dz) \cdot \boldsymbol{A}^* + \text{c.c.}]$$

$$(\text{C.3})$$

The Lagrangian corresponding to the second term of Equation (C.2) is

$$\mathscr{L}_2 = -(\omega/4cn)\,\boldsymbol{P} \cdot \boldsymbol{A}^*$$

$$(\text{C.4})$$

The nonlinear polarization amplitude is given by

$$\boldsymbol{P} = 3\chi^{(3)}(\boldsymbol{A} \cdot \boldsymbol{A}^*)\boldsymbol{A}$$

$$(\text{C.5})$$

in the case of an isotropic medium, where $\chi^{(3)}$ is the third-order nonlinearity. The pump, probe and phase-conjugate waves are identically polarized and the electric fields are taken as those in Equation (3.15). The electric fields are substituted in Equation (C.4) and only those terms that satisfy the phase-matching conditions are selected to obtain the Lagrangians for all groups:

$$\mathscr{L} = (1/2)[i(A_f{}^*A_f{}' - A_b{}^*A_b{}' + A_p{}^*A_p{}' - A_c{}^*A_c{}') + \text{c.c.}]$$
$$- (3\omega\chi^{(3)}/2cn)[(1/2)(|A_f|^2 + |A_b|^2 + |A_p|^2 + |A_c|^2)^2$$
$$+ |A_fA_b|^2 + |A_fA_p|^2 + |A_fA_c|^2 + |A_bA_p|^2 + |A_bA_c|^2 + |A_pA_c|^2$$
$$+ 2(A_fA_bA_p{}^*A_c{}^* + A_f{}^*A_b{}^*A_pA_c)]$$

$$(\text{C.6})$$

where $' = d/dz$. If a Legendre transformation

$$H = \sum_j p_j A_j' - \mathcal{L}, \qquad p_j = \partial \mathcal{L} / \partial A_j' \tag{C.7}$$

is performed on Equation (C.6), the Hamiltonian H can be obtained by the following equation:

$$H = (3\omega\chi^{(3)}/2cn)\,[\,(1/2)\,(|A_f|^2 + |A_b|^2 + |A_p|^2 + |A_c|^2)^2$$
$$+ |A_f A_b|^2 + |A_f A_p|^2 + |A_f A_c|^2 + |A_b A_p|^2 + |A_b A_c|^2 + |A_p A_c|^2$$
$$+ 2(A_f A_b A_p{}^* A_c{}^* + A_f{}^* A_b{}^* A_p A_c)\,] \tag{C.8}$$

This Hamiltonian is a constant since it does not explicitly contain z.

We consider here invariants corresponding to gauge transformations. If the phase transformation $\exp(i\lambda_j)\,A_j$ is performed on a complex electric field A_j, the current density J_m becomes[6]

$$J_m = -i \sum_j \lambda_j (\partial \mathcal{L} / \partial A_j')\, A_j \tag{C.9}$$

If there are nonzero phase changes for two of the electric field components, Equation (3.38) is obtained as the conservation quantity. If Equations (3.38) and (3.39) are applied to Equation (C.8), we obtain the result of Equation (3.41) where, in the present situation, the constant Γ_0 becomes

$$\Gamma_0 = (cn/6\omega\chi^{(3)})\,H - (1/8)\,(J_1{}^2 + J_2{}^2 + J_3{}^2)$$
$$- (1/2)\,(J_1 J_2 + J_2 J_3 + J_3 J_1) \tag{C.10}$$

From here the derivation proceeds in the same way as in Section 3.2(4).

In this connection, if we apply Equation (C.6) to the Lagrange-Euler equation

$$\sum_{m=1}^{4} \frac{\partial}{\partial x_m}\left(\frac{\partial \mathcal{L}}{\partial A_{j,m}}\right) = \frac{\partial \mathcal{L}}{\partial A_j} \tag{C.11}$$

we obtain the same coupling equation as Equation (3.20), where the x_m are the time-space coordinates, and the $A_{j,m}$ are the partial derivatives of the A_j with respect to the x_m.

D. Quantum Theory of Optically-Squeezed States and Degenerate Four-Wave Mixing (Addendum)

We will explain the theoretical substantiation for how the optically-squeezed state is produced and the quantum theory of degenerate four wave mixing. This Appendix is closely related to Sections 4.2 and 8.8.

(1) Theory of the Production of the Optically-Squeezed State

Assume the same degenerate four-wave mixing optical system as in Figure 3.1. We will take a classical approach since the two pump wave intensities are sufficiently strong so we can assume that depletion during propagation can be neglected. If the positions of the probe wave (subscript p) and the phase-conjugate wave (subscript c) are quantized, the complex amplitudes $A_j, A_j^*(j = p, c)$ of the electric field become the photon annihilation and creation operators $\hat{a}_j$ and $\hat{a}_j^{\dagger}$, respectively. The coupled-mode equations satisfied by the probe and phase-conjugate waves — already given by Equations (3.22a, b) — can be written as

$$\frac{d\hat{a}_p^{\dagger}}{dz} = ix\hat{a}_c \tag{D.1a}$$

$$\frac{d\hat{a}_c}{dz} = ix^*\hat{a}_p^{\dagger} \tag{D.1b}$$

where the coupling coefficient x is the same as that in Equation (3.23) and where each operator satisfies Transformation Function (8.32). The solutions of Equations (D.1a, b) can be obtained from

$$\hat{a}_c(0) = \hat{a}_c(L)\sec(xL) - i\hat{a}_p^{\dagger}(0)\tan(xL) \tag{D.2a}$$

$$\hat{a}_p(L) = \hat{a}_p(0)\sec(xL) - i\hat{a}_c^{\dagger}(L)\tan(xL) \tag{D.2b}$$

the forms of which are the same as in classical theory. However, there is a deviation from classical theory since the incident waves $\hat{a}_p(0)$ and $\hat{a}_c(L)$ always fluctuate so that we cannot assume that $\hat{a}_c(L) = 0$ in Equations (D.2a) and (D.2b).

In order to obtain the probe and phase-conjugate waves' quantum fluctuations, the orthogonal phase component can be described by

$$\hat{X}_{j1} \equiv (\hat{a}_j + \hat{a}_j^{\dagger})/2, \quad \hat{X}_{j2} \equiv (\hat{a}_j - \hat{a}_j^{\dagger})/2i \quad (j = p, c) \tag{D.3}$$

with the dispersion defined in the same way as Equation (8.34). The fluctuation of the incident wave is equivalent to that of the coherent state; that is, if we assume that

$$\langle(\Delta\hat{X}_{pm}(0))^2\rangle = \langle(\Delta\hat{X}_{cm}(L))^2\rangle = 1/4 \quad (m = 1, 2) \tag{D.4}$$

then we obtain

$$\langle(\Delta\hat{X}_{pm}(L))^2\rangle = \langle(\Delta\hat{X}_{cm}(0))^2\rangle$$
$$= [1 + 2\tan^2(xL)]/4 \tag{D.5}$$

as the exit wave's fluctuation. When we compare Equations (D.4) and (D.5), we see that the probe-wave fluctuation and the phase-conjugate wave fluctuation increase together as a function of four-wave mixing.

If a probe and phase-conjugate wave pass through a semitransparent mirror from two wave paths to produce the optically-squeezed state, since the phase is varied by the transmitted and reflected waves, the photon annihilation operator after the waves have passed through the semitransparent mirror can be described as follows:

$$\hat{b}_3 = [\hat{a}_c(0) - i\hat{a}_p(L)]/\sqrt{2} \tag{D.6a}$$

$$\hat{b}_4 = [\hat{a}_c(0) + i\hat{a}_p(L)]/\sqrt{2} \tag{D.6b}$$

where the position of the transmitted wave is taken as $\hat{b}_3$ ($\hat{b}_4$) for the probe (phase-conjugate) wave. Equations (D.2a, b) are substituted in the above equations and we obtain the following equations for the $\hat{b}_j$ ($j = 3, 4$) quantum fluctuations $<(\Delta\hat{Y}_{jm})^2>$:

$$\langle(\Delta\hat{Y}_{j1})^2\rangle = [\sec(\varkappa L) - \tan(\varkappa L)]^2/4 \tag{D.7a}$$

$$\langle(\Delta\hat{Y}_{j2})^2\rangle = [\sec(\varkappa L) + \tan(\varkappa L)]^2/4 \tag{D.7b}$$

Since the above equations imply $<(\Delta\hat{Y}_{j1})^2><(\Delta\hat{Y}_{j2})^2> = 1/16$, the combination of $\hat{b}_3$ and $\hat{b}_4$ is clearly the minimum uncertainty state. When $\varkappa L \to \pi/2$ in Equations (D.7), $<(\Delta\hat{Y}_{j1})^2> \to 0$. This means that the measured error becomes extremely small.

A wave injected into a nonlinear medium is used here and the annihilation operators are

$$\hat{c}_3 = [\hat{a}_c(L) - i\hat{a}_p(0)]/\sqrt{2} \tag{D.8a}$$

$$\hat{c}_4 = [\hat{a}_c(L) + i\hat{a}_p(0)]/\sqrt{2} \tag{D.8b}$$

If the parameters are defined by $\mu = \sec(\varkappa L)$ and $\nu = \tan(\varkappa L)$, $\hat{b}_3$ and $\hat{b}_4$ can be represented by

$$\hat{b}_j = \mu\hat{c}_j - \nu\hat{c}_j^{\dagger} \quad (j = 3, 4) \tag{D.9}$$

Here $\mu^2 - \nu^2 = 1$, which shows that $\hat{b}_3$ and $\hat{b}_4$ are in the optically-squeezed state. In short, according to the arrangement described above, this means that the optically-squeezed states $\hat{b}_3$ and $\hat{b}_4$ are produced from coherent states $\hat{a}_p(0)$ and $\hat{a}_c(L)$.

Although only the principle for production of the optical coherent state is explained above, it is also necessary to apply quantum theory (as in the following section) in order to investigate the effects of absorption, saturation and dispersion.

(2) Detailed Derivation of Quantum Theory of Degenerate Four-Wave Mixing (Addendum to Section 4.2)

The Hamiltonian for the complete system is obtained from the following equation:

$$H = \int \frac{d^3r}{\delta v} H_r \tag{D.10}$$

A dense-medium approximation is used for integrating the above equation.

We next derive the Fokker-Planck equation. We can write the following equation for the characteristic function F, taking $f(\eta)$ as the probability distribution function:

$$F = \mathrm{Tr}[\hat{O}(\boldsymbol{\lambda})\hat{\rho}(t)]$$
$$= \int f(\eta)\exp(i\boldsymbol{\lambda}\cdot\boldsymbol{\eta})\,d^2\eta_1\cdots d^2\eta_5 \tag{D.11}$$

where

$$\hat{O}(\boldsymbol{\lambda}) = \exp(i\lambda_5\hat{S}^\dagger)\exp(i\lambda_4\hat{S}_z)\exp(i\lambda_3\hat{S})\exp(i\lambda_2\hat{a}_r{}^\dagger)\exp(i\lambda_1\hat{a}_r) \tag{D.12a}$$

$$\boldsymbol{\lambda} = (\lambda_1,\ \lambda_2,\ \lambda_3,\ \lambda_4,\ \lambda_5) \tag{D.12b}$$

$$\boldsymbol{\eta} = (\eta_1,\ \eta_2,\ \eta_3,\ \eta_4,\ \eta_5) = (\alpha',\ \alpha'^\dagger,\ v',\ D,\ v'^\dagger) \tag{D.12c}$$

$$\hat{S} = \sum_{m=1}^{N_0} \hat{\sigma}_m = \sum_{m=1}^{N_0} (|1\rangle\langle 2|)_m,\quad \hat{S}_z = \sum_{m=1}^{N_0} \hat{\sigma}_{zm} = \hat{N}_2 - \hat{N}_1 \tag{D.12d}$$

$$\hat{N}_l \equiv \sum_{m=1}^{N_0} (|l\rangle\langle l|)_m \tag{D.12e}$$

Here, $\hat{O}(\lambda)$ is the normal ordering operator, $\hat{S}$ is the atomic dipole moment, $\hat{N}_l$ is the number of atoms in state l, and $\hat{S}_z$ is the difference in the atom numbers for state 2. In addition, corresponding functions for the c number and the operator are taken from the following:

$$v'^* \to \hat{S}^\dagger,\ \ D' \to \hat{S}_z,\ \ v' \to \hat{S},\ \ \alpha'^* \to \hat{a}_r{}^\dagger,\ \ \alpha' \to \hat{a}_r \tag{D.13}$$

On the other hand, the probability distribution function can also be written as

$$f(\eta) = \mathrm{Tr}[\hat{\rho}(t)\,\delta(v^* - \hat{S}^\dagger)\,\delta(D - \hat{S}_z)\,\delta(v - \hat{S})\,\delta(\alpha^* - \hat{a}_r{}^\dagger)\,\delta(\alpha - \hat{a}_r)] \tag{D.14}$$

where

$$\delta(\eta - \hat{O}) = \frac{1}{2\pi}\int_{-\infty}^{\infty}\exp[-i\lambda(\eta - \hat{O})]\,d\lambda$$
$$= \exp[-\hat{O}(\partial/\partial\eta)]\,\delta(\eta) \tag{D.15}$$

and δ represents the delta function. If we differentiate the distribution function with respect to t, we obtain

$$\partial f(\eta)/\partial t = \sum_{k=1}^{4} T_k \tag{D.16a}$$

$$T_k \equiv (-i/\hbar)\,\mathrm{Tr}\{\hat{\rho}[\prod_j \delta(\eta_j - \hat{O}_j),\ H_k]\}\quad (k=1,\ 2,\ 3) \tag{D.16b}$$

$$T_4 \equiv \mathrm{Tr}\{L_4[\hat{\rho}]\prod_j \delta(\eta_j - \hat{O}_j)\} \tag{D.16c}$$

After computing the T_k, we obtain an equation that contains an infinite number of derivatives. If we expand in terms of $1/N_0$ for sufficiently large N_0 and keep derivatives up to the second, we obtain the Fokker-Planck equation as follows:

$$\frac{df}{dt} = \left\{ -\frac{\partial}{\partial a}(gv) - \frac{\partial}{\partial v}(-\gamma_t v + gD\alpha) - \frac{\partial}{\partial D}[-\gamma_1(D+N_0) \right.$$

$$-2g(v^\dagger\alpha + v\alpha^\dagger)] + \frac{1}{2}\frac{\partial^2}{\partial v^2}(2gv\alpha) + \frac{1}{2}\frac{\partial^2}{\partial D^2}[2\gamma_1(D+N_0)$$

$$\left. -4g(v^\dagger\alpha + v\alpha^\dagger)] + \text{c.c.}\right\}f \tag{D.17}$$

where $\gamma_t = \gamma_{12}(1 + i\delta)$, and c.c. denotes the same expression with the substitution of $v^\dagger$ for v and $\alpha^\dagger$ for α.

The first derivative in the above equation is the drift term and the second derivative corresponds to the scattering term.

The following equations are obtained from Equation (D.17):

$$dv/dt = -\gamma_t v + gD\alpha + \Gamma_v \tag{D.18a}$$

$$dD/dt = -\gamma_1(D+N_0) - 2g(v^\dagger\alpha + v\alpha^\dagger) + \Gamma_D \tag{D.18b}$$

$$d\alpha/dt = gv + \Gamma_\alpha \tag{D.18c}$$

where Γ_j is a random variable such that $<\Gamma_j(t)\Gamma_j(t')> \propto \delta(t - t')$; this happens because of the dispersion term in Equation (D.17). This corresponds to the Langevin force. The equations for $v^\dagger$ and $\alpha^\dagger$ are (D.18a) and (D.18c), respectively, when v is replaced by $v^\dagger$, when α is replaced by $\alpha^\dagger$, and when the c number is obtained by complex conjugation. Now if we assume that the atomic variables relax faster than the wave position variables and that $dv/dt = 0$, $dv^\dagger/dt = 0$, and $dD/dt = 0$, then we can obtain the atom number difference D as well as the atomic polarization v. This technique is called the adiabatic method. The D and v in quantum theory correspond respectively to $(\rho_{22} - \rho_{11})$ and ρ_{12} in semiclassical theory.

Now the wave location α is separated into the sum ξ of the two high-intensity pump waves and the sum ζ of the probe and phase-conjugate waves. If we consider the probe and phase-conjugate waves up to first order in small quantities, we can obtain Equation (4.25) as the atomic polarization at position **r**.

We next derive the coupled-mode equations for probe and phase-conjugate waves. The Heisenberg equation of motion for probe and phase-conjugate waves is given by Equation (4.27). Therefore, the equation for the amplitude of a wave position is given as follows. If we assume cyclic averages of Π^{-1} and Π^{-2} in the integral on the right-hand side of Equation (4.28), we ultimately obtain the following coupled-mode equations for the probe and phase-conjugate waves:

$$d\alpha_p(t)/dt = -iK^*c\alpha_c^\dagger(t) - (a/2)c\alpha_p(t) + \Gamma_p(t) \tag{D.19a}$$

$$d\alpha_c(t)/dt = -iK^*c\alpha_p^\dagger(t) - (a/2)c\alpha_c(t) + \Gamma_c(t) \tag{D.19b}$$

where K and α are the coupling coefficient and the intensity absorption coefficient shown in Equations (4.30) and (4.31), respectively; and $\Gamma_j(t)$ is a random function. If these are shown as position variables, Equations (4.29a) and (4.29b) become the fundamental equations of quantum theory.

Bibliography

Chapter 1

1) B. Y. Zel'dovich, V. I. Popovichev, V. V. Ragul'skii and F. S. Faizullov: Connection between the wave fronts of the reflected and exciting light in stimulated Mandel'shtam-Brillouin scattering. *Sov. Phys. JETP Lett.*, **15** (1972), 109-113

2) B. I. Stepanov, E. V. Ivakin and A. S. Rubanov: Recording two-dimensional and three-dimensional dynamic holograms in transparent substances. *Sov. Phys. Doklady*, **16** (1971), 46-48

3) A. Yariv: On transmission and recovery of three-dimensional image information in optical waveguides. *J. Opt. Soc. Am.*, **66** (1976), 301-306

4) R. W. Hellwarth: Generation of time reversed wave fronts by nonlinear refraction. *J. Opt. Soc. Am.*, **67** (1977), 1-3

5) A. Yuriv and D. M. Pepper: Amplified reflection, phase conjugation and oscillation in degenerate four-wave mixing. *Opt. Lett.*, **1** (1977), 16-18

6) D. M. Bloom and G. C. Bjorklund: Conjugate wave-front generation and image reconstruction by four-wave mixing. *Appl. Phys. Lett.*, **31** (1977), 592-594

7) A. Ashkin, G. D. Boyd, J. M. Dziedzic, R. G. Smith, A. A. Ballman, J. J. Levinstein and K. Nassau: Optically-induced refractive index inhomogeneities in $LiNbO_3$ and $LiTaO_3$. *Appl. Phys. Lett.*, **9** (1966), 72-74

8) J. P. Huignard, J. P. Herriau, P. Aubourg and E. Spitz: Phase-conjugate wavefront generation via real-time holography in $Bi_{12}SiO_{20}$ crystals. *Opt. Lett.*, **4** (1979), 21-23

9) J. O. White, M. Cronin-Golomb, B. Fischer and A. Yariv: Coherent oscillation by self-induced gratings in the photorefractive crystal $BaTiO_3$. *Appl. Phys. Lett.*, **40** (1982), 450-452

10) V. Wang and C. R. Giuliano: Correction of phase aberrations via stimulated Brillouin scattering. *Opt. Lett.*, **2** (1978), 4-6

11) A. Yariv: Four wave nonlinear optical mixing as real time holography. *Opt. Commun.*, **25** (1978), 23-25

Chapter 2

1) A. Yariv: Phase conjugate optics and real-time holography. *IEEE J. Quantum Electron.*, **QE-14** (1978), 650-660; Comments and author's reply. **QE-15** (1979), 523-525

2) D. M. Pepper: Nonlinear optical phase conjugation. *Opt. Eng.*, **21** (1982), 156-183

3) C. J. Bordé: Phase conjugate optics and applications to interferometry and to laser gyros in "Quantum Optics, Experimental Gravitation and Measurement Theory" (ed. by P. Meystre and M. O. Scully), Plenum Press, New York, 1983, 269-291

References 1) and 2) are good commentaries that explain the basic concepts and applications of phase-conjugate optics. The special edition described below deals comprehensively with phase-conjugate optics. It contains fifteen monographs.

4) *Opt. Eng.*, vol. 21, no. 2 (1982)

The following separate volume by almost the same contributors has been published.

5) R. A. Fisher: "Optical Phase Conjugation," Academic Press, New York, 1983

Reference 5) has fourteen chapters. Although it is a good reference book that gives a fully comprehensive account of phase-conjugate optics, symbols are inconsistent due to the variety of contributors.

6) B. Y. Zel'dovich, N. F. Pilipetsky and V. V. Shkunov: "Principles of Phase Conjugation," Springer-Verlag, Berlin, 1985

Reference 6) is a literary masterpiece by Zel'dovich et al., the discoverers of phase-conjugate optics, and has a detailed description of stimulated scattering.

Chapter 3

1) R. W. Minck, R. W. Terhune and C. C. Wang: Nonlinear optics. *Appl. Opt.*, **5** (1966), 1595-1612

2) P. D. Maker and R. W. Terhune: Study of optical effects due to an induced polarization third order in the electric field strength. *Phys. Rev.*, **137** (1965), A801-A818

3) A. Yariv: On transmission and recovery of three-dimensional image information in optical waveguides. *J. Opt. Soc. Am.*, **66** (1976), 301-306

4) R. W. Hellwarth: Generation of time reversed wave fronts by nonlinear refraction. *J. Opt. Soc. Am.*, **67** (1977), 1-3

5) D. M. Pepper: Nonlinear optical phase conjugation. *Opt. Eng.*, **21** (1982), 156-183

6) A. Yariv and D. M. Pepper: Amplified reflection, phase conjugation and oscillation in degenerate four-wave mixing. *Opt. Lett.*, **1** (1977), 16-18

7) C. G. Caro and M. C. Gower: Phase conjugation by degenerate four-wave mixing in absorbing media. *IEEE J. Quantum Electron.*, **QE-18** (1982), 1376-1380

8) J. A. Armstrong, N. Bloembergen, J. Ducuing and P. S. Pershan: Interactions between light waves in a nonlinear dieletric. *Phys. Rev.*, **127** (1962), 1918-1939

9) J. H. Marburger and J. F. Lam: Nonlinear theory of degenerate four-wave mixing. *Appl. Phys. Lett.*, **34** (1979), 389-391

10) C. Paré, M. Piché and P. Bélanger: Stability analysis and temporal evolution of

degenerate four-wave mixing and phase-conjugate lasers. *J. Opt. Soc. Am. B*, **5** (1988), 679-689

11) D. M. Pepper and R. L. Abrams: Narrow optical bandpass filter via nearly degenerate four-wave mixing. *Opt. Lett.*, **3** (1978), 212-214

12) N. V. Kukhtarev, V. B. Markov, S. G. Odulov, M. S. Soskin and V. L. Vinetskii: Holographic storage in electrooptic crystals. I. Steady state. *Ferroelectrics*, **22** (1979), 949-960

13) J. Feinberg, D. Heiman, A. R. Tanguay, Jr. and R. W. Hellwarth: Photorefractive effects and light-induced charge migration in barium titanate. *J. Appl. Phys.*, **51** (1980), 1297-1305; Erratum, **52** (1981), 537

14) R. A. Mullen and R. W. Hellwarth: Optical measurement of the photorefractive parameters of $Bi_{12}SiO_{20}$. *J. Appl. Phys.*, **58** (1985), 40-44

15) S. Ducharme and J. Feinberg: Altering the photorefractive properties of $BaTiO_3$ by reduction and oxidation at 650°C. *J. Opt. Soc. Am. B*, **3** (1986), 283-292

16) M. Cronin-Golomb, B. Fischer, J. O. White and A. Yariv: Theory and applications of four-wave mixing in photorefractive media. *IEEE J. Quantum Electron.*, **QE-20** (1984), 12-30

17) D. L. Staebler and J. J. Amodei: Coupled-wave analysis of holographic storage in $LiNbO_3$. *J. Appl. Phys.*, **43** (1972) 1042-1049

18) B. Fischer, M. Cronin-Golomb, J. O. White and A. Yariv: Amplified reflection, transmission and self-oscillation in real-time holography. *Opt. Lett.*, **6** (1981), 519-521

19) H. Kogelnik: Coupled wave theory for thick hologram gratings. *Bell System Tech. J.*, **48** (1969), 2909-2947

20) S. I. Stepanov and M. P. Petrov: Degenerate four-wave mixing via shifted phase holograms in cubic photorefractive crystals. *Opt. Commun.*, **53** (1985), 64-68

21) S. I. Stepanov and M. P. Petrov: Nonstationary holographic recording for efficient amplification and phase conjugation in "Photorefractive materials and their applications I: Fundamental phenomena" (ed. by P. Günter and J. P. Huignard), Springer-Verlag, Berlin, 1988, 263-289

22) Yasuhine Doi: "Polarization and Crystal Optics," Kyoritsu Publishing Company, 1980, Chapter 2.

23) B. Fischer and S. Weiss: Solvable optimized four-wave mixing configuration with cubic photorefractive crystals. *Appl. Phys. Lett.*, **53** (1988), 257-259

24) M. R. Belić: Exact solution to the degenerate four-wave mixing in reflection geometry in photorefractive media. *Phys. Rev. A*, **31** (1985), 3169-3174

25) M. R. Belić and M. Lax: Exact solution to the stationary holographic four-wave mixing in photorefractive crystals. *Opt. Commun.*, **56** (1985), 197-201

26) M. Cronin-Golomb, J. O. White, B. Fischer and A. Yariv: Exact solution of a nonlinear model of four-wave mixing and phase conjugation. *Opt. Lett.*, **7** (1982), 313-315

27) W. Królikowski, M. R. Belić and A. Bledowski: Phase transfer in optical phase conjugation. *Phys. Rev. A*, **37** (1988), 2224-2226

28) N. V. Kukhtarev: Kinetics of hologram recording and erasure in electrooptic crystals. *Sov. Tech. Phys. Lett.*, **2** (1976), 438-440

29) J. P. Huignard, H. Rajbenbach, P. Refrégier and L. Solymar: Wave mixing in photorefractive bismuth silicon oxide crystals and its applications. *Opt. Eng.*, **24** (1985) 586-592

30) P. Refrégier, L. Solymar, H. Rajbenbach and J. P. Huignard: Two-beam coupling in photorefractive $Bi_{12}SiO_{20}$ crystals with moving grating: Theory and experiments. *J. Appl. Phys.*, **58** (1985), 45-57

31) G. C. Valley: Two-wave mixing with an applied field and a moving grating. *J. Opt. Soc. Am. B*, **1** (1984), 868-873

32) J. F. Lam: Spectral response of nearly degenerate four-wave mixing in photo-refractive materials. *Appl. Phys. Lett.*, **42** (1983), 155-157

33) J. Sakai: Polarization dependence of reflectivity of phase-conjugate wave generated by degenerate four-wave mixing in $Bi_{12}SiO_{20}$. *Jpn. J. Appl. Phys.*, **27** (1988), 1899-1905

34) A. Erdmann and R. Kowarschik: Theory of degenerate four-wave mixing in anisotropic, optically active photorefractive crystals. *IEEE J. Quantum Electron.*, **24** (1988), 155-161

35) A. Yariv: Four wave nonlinear optical mixing as real time holography. *Opt. Commun.*, **25** (1978), 23-25

Chapter 4

1) R. L. Abrams, J. F. Lam, R. C. Lind, D. G. Steel and P. F. Liao: Phase conjugation and high-resolution spectroscopy by resonant degenerate four-wave mixing in "Optical phase conjugation" (ed. by R. A. Fisher), Academic Press, New York, 1983, 211-284

2) M. Sargent III, M. O. Scully and W. E. Lamb, Jr.: "Laser Physics" [Translated by Shimoda, Iwasawa, and Kamiya (Into Japanese)], Maruzen, 1978, Chapters 7 and 8.

3) R. L. Abrams and R. C. Lind: Degenerate four-wave mixing in absorbing media. *Opt. Lett.*, **2** (1978), 94-96: Errata, **3** (1978), 205

4) R. C. Lind, D. G. Steel and G. J. Dunning: Phase conjugation by resonantly enhanced degenerate four-wave mixing. *Opt. Eng.*, **21** (1982), 190-198

5) W. P. Brown: Absorption and depletion effects on degenerate four-wave mixing in homogeneously broadened absorbers. *J. Opt. Soc. Am.*, **73** (1983), 629-634

6) T. Fu and M. Sargent III: Effects of signal detuning on phase conjugation. *Opt. Lett.*, **4** (1979), 366-368

7) S. M. Wandzura: Effects of atomic motion on wavefront conjugation by resonantly enhanced degenerate four-wave mixing. *Opt. Lett.*, **4** (1979), 208-210

8) G. Grynberg, B. Kleinmann, M. Pinard and P. Verkerk: Amplified reflections in degenerate four-wave mixing: a more-accurate theory. *Opt. Lett.*, **8** (1983), 614-616

9) H. Haken, H. Risken and W. Weidlich: Quantum mechanical solutions of the laser masterequation III. Exact equation for a distribution function of macroscopic variables. *Zeitschrift für Physik*, **206** (1967), 355-368

10) W. H. Louisell: "Quantum Statistical Properties of Radiation," John Wiley & Sons, New York, 1973

11) M. D. Reid and D. F. Walls: Generation of squeezed states via degenerate four-wave mixing. *Phys. Rev. A*, **31** (1985), 1622-1635

12) R. S. Bondurant, P. Kumar, J. H. Shapiro and M. Maeda: Degenerate four-wave mixing as a possible source of squeezed-state light. *Phys. Rev. A*, **30** (1984) , 343-353

13) J. F. Lam: Doppler-free laser spectroscopy via degenerate four-wave mixing. *Opt. Eng.*, **21** (1982), 219-223

14) A. Maruani: Propagation analysis of forward degenerate four-wave mixing. *IEEE J. Quantum Electron.* **QE-16** (1980), 558-566

15) A. Khyzniak, V. Kondilenko, Y. Kucherov, S. Lesnik, S. Odulov and M. Soskin: Phase conjugation by degenerate forward four-wave mixing. *J. Opt. Soc. Am. A*, **1** (1984), 169-175

16) C. V. Heer and P. F. McManamon: Wavefront correction with photon echoes. *Opt. Commun.*, **23** (1977), 49-50

17) Koichi Shimoda, "Introduction to Laser Physics," Iwanami Booksellers, 1987, 170-172

18) J. C. AuYeung: Phase conjugation from nonlinear photon echoes in "Optical Phase Conjugation" (ed. by R. A. Fisher), Academic Press, New York, 1983, 285-305

19) A. Korpel and M. Chatterjee: Nonlinear echoes, phase conjugation, time reversal and electronic holography. *Proc. IEEE*, **69** (1981), 1539-1556

20) J. G. Fujimoto and T. K. Yee: Diagrammatic density matrix theory of transient four-wave mixing and the measurement of transient phenomena. *IEEE J. Quantum Electron.* **QE-22** (1986), 1215-1228

21) A. M. Weiner, S. De Silvestri and E. P. Ippen: Three-pulse scattering for femtosecond dephasing studies: theory and experiment. *J. Opt. Soc. Am. B*, **2** (1985), 654-661

22) T. Yajima and Y. Taira: Spatial optical parametric coupling of picosecond light pulses and transverse relaxation effect in resonant media. *J. Phys. Soc. Jpn.*, **47** (1979), 1620-1626

23) R. A. Fisher, B. R. Suydam and B. J. Feldman: Transient analysis of Kerr-like phase conjugators using frequency-domain techniques. *Phys. Rev. A*, **23** (1981), 3071-3083

24) J. H. Marburger: Optical pulse integration and chirp reversal in degenerate four-wave mixing. *Appl. Phys. Lett.*, **32** (1978), 372-374

25) G. C. Papen, B. E. A. Saleh and J. Tataronis: Analysis of transient phase conjugation in photorefractive media. *J. Opt. Soc. Am. B*, **5** (1988), 1763-1774

Chapter 5

1) P. V. Avizonis, F. A. Hopf, W. D. Bomberger, S. F. Jacobs, A. Tomita and K. H. Womack: Optical phase conjugation in a lithium formate crystal. *Appl. Phys. Lett.*, **31** (1977), 435-437

2) B. Y. Zel'dovich, V. I. Popovichev, V. V. Ragul'skii and F. S. Faizullov: Connection between the wave fronts of the reflected and exciting light in stimulated Mandel'shtam-Brillouin scattering. *Sov. Phys. JETP Lett.*, **15** (1972), 109-113

3) D. Cotter: Stimulated Brillouin scattering in monomode optical fiber. *J. Opt. Commun.*, **4** (1983), 10-19

4) R. W. Hellwarth: Theory of phase conjugation by stimulated scattering in a waveguide. *J. Opt. Soc. Am.*, **68** (1978), 1050-1056

5) B. Y. Zel'dovich and V. V. Shkunov: Wavefront reproduction in stimulated Raman scattering. *Sov. J. Quantum Electron.*, **7** (1977) 610-615

Chapter 6

1) A. Yariv, J. AuYeung, D. Fekete and D. M. Pepper: Image phase compensation and real-time holography by four-wave mixing in optical fibers. *Appl. Phys. Lett.*, **32** (1978), 635-637

2) D. M. Bloom, P. F. Liao and N. P. Economou: Observation of amplified reflection by degenerate four-wave mixing in atomic sodium vapor. *Opt. Lett.*, **2** (1978), 58-60

3) R. H. Stolen and A. Ashkin: Optical Kerr effect in glass waveguide. *Appl. Phys. Lett.*, **22** (1973), 294-296

4) J. AuYeung, D. Fekete, D. M. Pepper, A. Yariv and R. K. Jain: Continuous backward-wave generation by degenerate four-wave mixing in optical fibers. *Opt. Lett.*, **4** (1979), 42-44

5) A. Gabel, K. W. Delong, C. T. Seaton and G. I. Stegeman: Efficient degenerate four-wave mixing in an ion-exchanged semiconductor-doped glass waveguide. *Appl. Phys. Lett.*, **51** (1987), 1682-1684

6) D. M. Pepper, D. Fekete and A. Yariv: Observation of amplified phase-conjugate reflection and optical parametric oscillation by degenerate four-wave mixing in a transparent medium. *Appl. Phys. Lett.*, **33** (1978), 41-44

7) R. K. Jain and M. B. Klein: Degenerate four-wave mixing in semiconductors in "Optical phase conjugation" (ed. by R. A. Fisher), Academic Press, New York, 1983, p. 314

8) S. C. Rand: Ultraviolet four-wave mixing and phase conjugation with metastable color-center states in diamond. *Opt. Lett.*, **13** (1988), 140-142

9) M. N. Islam, E. P. Ippen, E. G. Burkhardt and T. J. Bridges: Picosecond study of near-band-gap nonlinearities in GaInAsP. *J. Appl. Phys.*, **59** (1986), 2619-2628

10) A. E. Neeves and M. H. Birnboim: Polarization selective optical phase conjugation in a Kerr-like medium. *J. Opt. Soc. Am. B*, **5** (1988), 701-708

11) R. G. Caro and M. C. Gower: Phase conjugation of KrF laser radiation. *Opt. Lett.*, **6** (1981), 557-559

12) B. J. Feldman, R. A. Fisher and S. L. Shapiro: Ultraviolet phase conjugation. *Opt. Lett.*, **6** (1981), 84-86

13) Y. Silberberg and I. Bar-Joseph: Low power phase conjugation in thin films of saturable absorbers. *Opt. Commun.*, **39** (1981), 265-268

14) P. W. Smith, A. Ashkin and W. J. Tomlinson: Four-wave mixing in an artificial Kerr medium. *Opt. Lett.*, **6** (1981), 284-286

15) W. M. Dennis, W. Blau and D. J. Bradley: Optical phase conjugation in a soluble polymer. *Opt. Eng.*, **25** (1986), 538-540

16) R. K. Jain and R. C. Lind: Degenerate four-wave mixing in semiconductor-doped glasses. *J. Opt. Soc. Am.*, **73** (1983), 647-653

17) S. Mailhot, P. Galarneau, R. A. Lessard and M. Denariez-Roberge: Degenerate four-wave mixing in organic azo dyes chrysoidin and benzopurpurin 4B. *Appl. Opt.*, **27** (1988), 3418-3421

18) W. Richardson, L. Maleki and E. Garmire: Degenerate four-wave mixing in a mercury-argon discharge. *Opt. Lett.*, **11** (1986), 572-574

19) R. C. Lind and D. G. Steel: Demonstration of the longitudinal modes and aberration-correction properties of a continuous-wave dye laser with a phase-conjugate mirror. *Opt. Lett.*, **6** (1981), 554-556

20) J. O. Tocho, W. Sibbett and D. J. Bradley: Picosecond phase-conjugate reflection from organic dye saturable absorbers. *Opt. Commun.*, **34** (1980), 122-126

21) D. Fekete, J. AuYeung and A. Yariv: Phase-conjugate reflection by degenerate four-wave mixing in a nematic liquid crystal in the isotropic phase. *Opt. Lett.*, **5** (1980), 51-53

22) A. V. Mamaev, N. A. Melnikov, N. P. Pilipetsky, A. N. Sudarkin and V. V. Shkunov: Phase conjugation on the surface of semiconductors: theory and experiment. *J. Opt. Soc. Am. A*, **1** (1984), 856-859

23) D. Grischkowsky, N. S. Shiren and R. J. Bennett: Generation of time-reversed wave fronts using a resonantly enhanced electronic nonlinearity. *Appl. Phys. Lett.*, **33** (1978), 805-807

24) R. K. Jain, M. B. Klein and R. C. Lind: High-efficiency degenerate four-wave mixing of 1.06-μm radiation in silicon. *Opt. Lett.*, **4** (1979), 328-330

25) H. J. Eichler, J. Chen and K. Richter: Four-wave mixing reflectivity of silicon at 1.06 μm: Influence of free-carrier absorption. *Appl. Phys. B*, **42** (1987), 215-219

26) D. E. Watkins, C. R. Phipps, Jr. and S. J. Thomas: Observation of amplified reflection through degenerate four-wave mixing at CO_2 laser wavelengths in germanium. *Opt. Lett.*, **6** (1981), 76-78

27) D. E. Watkins, J. P. Figueira and S. J. Thomas: Observation of resonantly enhanced degenerate four-wave mixing in doped alkari halides. *Opt. Lett.*, **5** (1980), 169-171

28) P. A. Wolff, S. Y. Yuen, K. A. Harris, J. W. Cook, Jr. and J. F. Schetzina: Optical nonlinearity in mercury telluride. *Appl. Phys. Lett.*, **50** (1987), 1858-1860

29) R. K. Jain and D. G. Steel: Degenerate four-wave mixing of 10.6-μm radiation in $Hg_{1-x}Cd_xTe$. *Appl. Phys. Lett.*, **37** (1980), 1-3

30) R. C. Lind, D. G. Steel, M. B. Klein, R. L. Abrams, C. R. Giuliano and R. K. Jain: Phase conjugation at 10.6 μm by resonantly enhanced degenerate four-wave mixing. *Appl. Phys. Lett.*, **34** (1979), 457-459

31) S. Y. Yuen and P. A. Wolff: Difference-frequency variation of the free-carrier-induced, third-order nonlinear susceptibility in *n*-InSb. *Appl. Phys. Lett.*, **40** (1982), 457-459

32) J. Feinberg and R. W. Hellwarth: Phase-conjugating mirror with continous-wave gain. *Opt. Lett.*, **5** (1980), 519-521: Errata, **6** (1981), 257

33) H. Rajbenbach, J. P. Huignard and P. Refrégier: Amplified phase-conjugate beam reflection by four-wave mixing with photorefractive $Bi_{12}SiO_{20}$ crystals. *Opt. Lett.*, **9** (1984), 558-560

34) H. Rajbenbach, B. Imbert, J. P. Huignard and S. Mallick: Near-infrared four-wave mixing with gain and self-starting oscillators with photorefractive GaAs. *Opt. Lett.*, **14** (1989), 78-80

35) N. Tan-no, T. Hoshimiya and H. Inaba: Dispersion-free amplification and oscillation in phase-conjugate four-wave mixing in an atomic vapor doublet. *IEEE J. Quantum Electron.*, **QE-16** (1980), 147-153

36) D. M. Bloom and G. C. Bjorklund: Conjugate wave-front generation and image reconstruction by four-wave mixing. *Appl. Phys. Lett.*, **31** (1977), 592-594

37) R. K. Jain and M. B. Klein: Degenerate four-wave mixing near the band gap of semiconductors. *Appl. Phys. Lett.*, **35** (1979), 454-456

38) S. Y. Yuen: Degenerate four-wave mixing due to intervalence band transition in p-type mercury cadmium telluride. *Appl. Phys. Lett.*, **43** (1983), 479-481

39) P. A. Wolff and S. Y. Yuen: Nonlinear optics near the metal insulator transition. *Solid State Commun.*, **60** (1986), 645-648

40) R. K. Jain and G. J. Dunning: Spatial and temporal properties of a continous-wave phase-conjugate resonator based on the photorefractive crystal $BaTiO_3$. *Opt. Lett.*, **7** (1982), 420-422

41) J. O. White, M. Cronin-Golomb, B. Fischer and A. Yariv: Coherent oscillation by self-induced gratings in the photorefractive crystal $BaTiO_3$. *Appl. Phys. Lett.*, **40** (1982) 450-452

42) J. Feinberg: Self-pumped, continuous-wave phase conjugator using internal reflection. *Opt. Lett.*, **7** (1982), 486-488

43) A. M. Scott and P. Waggott: Phase conjugation by self-pumped Brillouin-induced four-wave mixing. *Opt. Lett.*, **12** (1987), 835-837

44) G. A. Rakuljic, K. Sayano and A. Yariv: Self-starting passive phase conjugate mirror with Ce-doped strontium barium niobate. *Appl. Phys. Lett.*, **50** (1987), 10-12

45) J. P. Huignard, J. P. Herriau, P. Aubourg and E. Spitz: Phase-conjugate wavefront generation via real-time holography in $Bi_{12}SiO_{20}$ crystals. *Opt. Lett.*, **4** (1979), 21-23

46) D. N. Rao, P. Chopra, S. K. Ghoshal, J. Swiatkiewicz and P. N. Prasad: Third-order nonlinear optical interaction and conformal transition in poly-4-BCMU polydi-

acetylene studied by picosecond and subpicosecond degenerate four wave mixing. *J. Chem. Phys.*, **84** (1986), 7049-7050

47) L. L. Chase, M. L. Claude, D. Hulin and A. Mysyrowicz: Optical phase conjugation based on the giant two-photon resonance of the excitonic molecule in CuCl. *Phys. Rev. A*, **28** (1983), 3696-3698

48) L. Yang, Q. Z. Wang, P. P. Ho, R. Dorsinville, R. R. Alfano, W. K. Zou and N. L. Yang: Ultrafast time response of optical nonlinearity in polysilane polymers. *Appl. Phys. Lett.*, **53** (1988), 1245-1247

49) G. M. Carter, M. K. Thakur, Y. J. Chen and J. V. Hryniewicz: Time and wavelength resolved nonlinear optical spectroscopy of a polydiacetylene in the solid state using picosecond dye laser pulses. *Appl. Phys. Lett.*, **47** (1985), 457-459

50) P. G. Huggard, W. Blau and D. Schweitzer: Large third-order optical nonlinearity of the organic metal α-[bis (ethylenedithio) tetrathiofulvalene] yriiodide. *Appl. Phys. Lett.*, **26** (1987), 2183-2185

51) D. A. B. Miller, D. S. Chemla, D. J. Eilenberger, P. W. Smith, A. C. Gossard and W. Wiegmann: Degenerate four-wave mixing in room-temperature GaAs/GaAlAs multiple quantum well structures. *Appl. Phys. Lett.*, **42** (1983), 925-927

52) D. S. Chemla, D. A. B. Miller, P. W. Smith, A. C. Gossard and W. Wiegmann: Room temperature excitonic nonlinear absorption and refraction in GaAs/AlGaAs multiple quantum well structures. *IEEE J. Quantum Electron.*, **QE-20** (1984), 265-275

53) G. Mizutani and N. Nagasawa: Phase-conjugation by four-wave mixing in CuCl: Resonance enhancement by exciton and excitonic molecule and its polarization dependence. *J. Phys. Soc. Jpn.*, **52** (1983), 2251-2257

54) D. G. Steel and J. F. Lam: Two-photon coherent-transient measurement of the nonradiative collisionless dephasing rate in SF_6 via Doppler-free degenerate four-wave mixing. *Phys. Rev. Lett.*, **43** (1979), 1588-1591

55) D. Bloch, M. Ducloy and E. Giacobino: Heterodyne Doppler-free two-photon absorption and dispersion in rubidium vapour. *J. Phys. B: At. Mol. Phys.*, **14** (1981), L 819-L 825

56) N. C. Griffen and C. V. Herr: Focusing and phase conjugation of photon echoes in Na vapor. *Appl. Phys. Lett.*, **33** (1978), 865-866

57) N. W. Carlson, W. R. Babbitt and T. W. Mossberg: Storage and phase conjugation of light pulses using stimulated photon echoes. *Opt. Lett.*, **8** (1983), 623-625

58) M. K. Kim and R. Kachru: Long-term image storage and phase conjugation by a backward-stimulated echo in Pr^{3+}: LaF_3. *J. Opt. Soc. Am. B*, **4** (1987), 305-308

59) G. M. Carter, J. V. Hryniewicz, M. K. Thakur, Y. J. Chen and S. E. Meyler: Nonlinear optical processes in a polydiacetylene measured with femtosecond duration laser pulses. *Appl. Phys. Lett.*, **49** (1986), 998-1000

60) J. L. Oudar, I. Abram and C. Minot: Picosecond degenerate four-wave mixing through orientation and concentration gratings in GaAs. *Appl. Phys. Lett.*, **44** (1984), 689-691

61) A. L. Smirl, T. F. Boggess and F. A. Hopf: Generation of a forward-traveling phase-conjugate wave in germanium. *Opt. Commun.*, **34** (1980), 463-468

62) A. L. Smirl, G. C. Valley, R. A. Mullen, K. Bohnert, C. D. Mire and T. F. Goggess: Picosecond photorefractive effect in BaTiO$_3$. *Opt. Lett.*, **12** (1987), 501-503

63) G. Lesaux, J. C. Launay and A. Brun: Transient photocurrent induced by nanosecond light pulses in BSO and BGO. *Opt. Commun.*, **57** (1986), 166-170

64) P. W. Smith, W. J. Tomlinson, D. J. Eilenberger and P. J. Maloney: Measurement of electronic optical Kerr coefficients. *Opt. Lett.*, **6** (1981), 581-583

65) J. M. Nunzi and D. Grec: Picosecond phase conjugation in polydiacetylene gels. *J. Appl. Phys.*, **62** (1987), 2198-2202

66) G. C. Valley, A. L. Smirl and M. B. Klein: Picosecond photorefractive beam coupling in GaAs. *Opt. Lett.*, **11** (1986), 647-649

67) P. F. Liao, N. P. Economou and R. R. Freeman: Two-photon coherent transient measurements of Doppler-free linewidths with broadband excitation. *Phys. Rev. Lett.*, **39** (1977), 1473-1476

68) P. V. Avizonis, F. A. Hopf, W. D. Bomberger, S. F. Jacobs, A. Tomita and K. H. Womack: Optical phase conjugation in a lithium formate crystal. *Appl. Phys. Lett.*, **31** (1977), 435-437

69) M. Slatkine, I. J. Bigio, B. J. Feldman and R. A. Fisher: Efficient phase conjugation of an ultraviolet XeF laser beam by stimulated Brillouin scattering. *Opt. Lett.*, **7** (1982), 108-110

70) B. Y. Zel'dovich, V. I. Popovichev, V. V. Ragul'skii and F. S. Faizullov: Connection between the wave fronts of the reflected and exciting light in stimulated Mandel'shtam-Brillouin scattering. *Sov. Phys. JETP Lett.*, **15** (1972) 109-113

71) V. Wang and C. R. Giuliano: Correction of phase aberrations via stimulated Brillouin scattering. *Opt. Lett.*, **2** (1978), 4-6

72) O. Y. Nosach, V. I. Popovichev, V. V. Ragul'skii and F. S. Faizullov: Cancellation of phase distortions in an amplifying medium with a "Brillouin mirror." *Sov. Phys. JETP Lett.*, **16** (1972), 435-438

73) A. A. Ilyukhin, G. V. Peregudov, M. E. Plotkin, E. N. Ragozin and V. A. Chirkov: Focusing of a laser beam on a target using the effect of wavefront inversion (WFI) produced as a result of stimulated Mandel'shtam-Brillouin scattering (SMBS). *JETP Lett.*, **29** (1979), 328-332

74) M. T. Duignan, B. J. Feldman and W. T. Whitney: Stimulated Brillouin scattering and phase conjugation of hydrogen flouride laser radiation. *Opt. Lett.*, **12** (1987), 111-113

75) R. Mays, Jr. and R. J. Lysiak: Phase conjugated wavefronts by stimulated Brillouin and Raman scattering. *Opt. Commun.*, **31** (1979), 89-92

76) B. Y. Zel'dovich, N. A. Mel'nikov, N. F. Pilipetskii and V. V. Ragul'skii: Observation of wave-front inversion in stimulated Raman scattering of light. *JETP Lett.*, **25** (1977), 36-39

77) A. M. Scott and K. D. Ridley: A review of Brillouin-enhanced four-wave mixing. *IEEE J. Quantum Electron.*, **25** (1989), 438-459

78) N. F. Andreev, V. I. Bespalov, A. M. Kiselev, A. Z. Matveev, G. A. Pasmanik and A. A. Shilov: Wave-front inversion of weak optical signals with a large reflection coefficient. *JETP Lett.*, **32** (1981), 625-629

Chapter 7

1) O. Svelto: Self-focusing, self-trapping and self-phase modulation of laser beams in "Progress in Optics" (ed. by E. Wolf), vol. XII, North-Holland, Netherlands, 1974, 3-51

2) R. W. Hellwarth: Third-order optical susceptibilities of liquids and solids. *Prog. Quantum Electron.*, **5** (1977) 1-68

3) R. W. Minck, R. W. Terhune and C. C. Wang: Nonlinear optics. *Appl. Opt.*, **5** (1966), 1595-1612

4) T. Y. Chang: Fast self-induced refractive index changes in optical media: a survey. *Opt. Eng.*, **20** (1981), 220-232

5) A. Javan and P. L. Kelly: Possibility of self-focusing due to intensity dependent anomalous dispersion. *IEEE J. Quantum Electron.*, **QE-2** (1966), 470-473

6) D. Grischkowsky and J. A. Armstong: Self-defocusing of light by adiabatic following in rubidium vapor. *Phys. Rev. A*, **6** (1972), 1566-1570

7) H. Eichler, G. Salje and H. Stahl: Thermal diffusion measurements using spatially periodic temperature distributions induced by laser light. *J. Appl. Phys.*, **44** (1973) 5383-5388

8) G. Martin and R. W. Hellwarth: Infrared-to-optical image conversion by Bragg reflection from thermally induced index gratings. *Appl. Phys. Lett.*, **34** (1979), 371-373

9) H. J. Hoffman: Thermally induced phase conjugation by transient real-time holography: a review. *J. Opt. Soc. Am. B*, **3** (1986), 253-273

10) Laser Optics Society: "Laser Handbook." Omu Company, 1982, p. 157.

11) A. Ashkin: Acceleration and trapping of particles by radiation pressure. *Phys. Rev. Lett.*, **24** (1970), 156-159

12) A. E. Neeves and M. H. Brinboim: Polarization selective optical phase conjugation in a Kerr-like medium. *J. Opt. Soc. Am. B*, **5** (1988), 701-708

13) P. W. Smith, A. Ashkin and W. J. Tomlinson: Four-wave mixing in an artificial Kerr medium. *Opt. Lett.*, **6** (1981), 284-286

14) A. Ashkin, G. D. Boyd, J. M. Dziedzic, R. G. Smith, A. A. Ballman, J. J. Levinstein and K. Nassau: Optically-induced refractive index inhomogeneities in $LiNbO_3$ and $LiTaO_3$. *Appl. Phys. Lett.* **9** (1966), 72-74

15) A. M. Glass: The photorefractive effect. *Opt. Eng.*, **17** (1978), 470-479

16) M. B. Klein and R. N. Schwartz: Photorefractive effect in $BaTiO_3$: microscopic origins. *J. Opt. Soc. Am. B*, **3** (1986), 293-305

17) M. Carrascosa and F. Agulló-López: Kinetics for optical erasure of sinusoidal holographic gratings in photorefractive materials. *IEEE J. Quantum Electron.*, **QE-22** (1986), 1369-1375

18) P. Günter and J. P. Huignard: Photorefractive effects and materials in "Photorefractive materials and their applications I: Fundamental phenomena" (ed. by P. Günter and J. P. Huignard), Springer-Verlag, Berlin, 1988, p. 51

19) G. C. Valley and M. B. Klein: Optimal properties of photorefractive materials for optical data processing. *Opt. Eng.*, **22** (1983), 704-711

20) M. D. Ewbank, R. R. Neurgaonkar, W. K. Cory and J. Feinberg: Photorefractive properties of strontium-barium niobate. *J. Appl. Phys.*, **62** (1987), 374-380

21) G. C. Valley: Two-wave mixing with an applied field and a moving grating. *J. Opt. Soc. Am. B*, **1** (1984), 868-873

22) J. C. Fabre, J. M. C. Jonathan and G. Roosen: Photorefractive beam coupling in GaAs and InP generated by nanosecond light pulses. *J. Opt. Soc. Am. B*, **5** (1988), 1730-1736

23) R. B. Bylsma, P. M. Bridenbaugh, D. H. Olson and A. M. Glass: Photorefractive properties of doped cadmium telluride. *Appl. Phys. Lett.*, **51** (1987) 889-891

24) S. Ducharme and J. Feinberg: Altering the photorefractive properties of $BaTiO_3$ by reduction and oxidation at 650°C. *J. Opt. Soc. Am. B*, **3** (1986), 283-292

25) P. A. Wolff, S. Y. Yuen, K. A. Harris, J. W. Cook, Jr. and J. F. Schetzina: Optical nonlinearity in mercury telluride. *Appl. Phys. Lett.*, **50** (1987), 1858-1860

26) R. K. Jain: Degenerate four-wave mixing in semiconductors: application to phase conjugation and to picosecond-resolved studies of transient carrier dynamics. *Opt. Eng.*, **21** (1982), 199-218

27) P. A. Wolff and G. A. Pearson: Theory of optical mixing by mobile carriers in semiconductors. *Phys. Rev. Lett.*, **17** (1966), 1015-1017

28) S. Y. Yuen and P. A. Wolff: Difference-frequency variation of the free-carrier-induced, third-order nonlinear susceptibility in n-InSb. *Appl. Phys. Lett.*, **40** (1982), 457-459

29) B. S. Wherrett and N. A. Higgins: Theory of nonlinear refraction near the band edge of a semiconductor. *Proc. R. Soc. London*, **A379** (1982), 67-90

30) D. S. Chemla, D. A. B. Miller, P. W. Smith, A. C. Gossard and W. Wiegmann: Room temperature excitonic nonlinear absorption and refraction in GaAs/GaAlAs multiple quantum well structures. *IEEE J. Quantum Electron.*, **QE-20** (1984), 265-275

31) A. M. Fox, A. C. Maciel, M. G. Shorthose, J. F. Ryan, M. D. Scott, J. I. Davies and J. R. Riffat: Nonlinear excitonic optical absorption in GaInAs/InP quantum wells. *Appl. Phys. Lett.*, **51** (1987), 30-32

32) J. Strait and A. M. Glass: Time-resolved photorefractive four-wave mixing in semiconductor materials. *J. Opt. Soc. Am B*, **3** (1986), 342-344

33) L. Cheng and A. Partovi: Index grating lifetime in photorefractive GaAs. *Appl. Opt.*, **27** (1988), 1760-1763

34) R. K. Jain and R. C. Lind: Degenerate four-wave mixing in semiconductor-doped glasses. *J. Opt. Soc. Am.*, **73** (1983), 647-653

35) N. F. Borrelli, D. W. Hall, H. J. Holland and D. W. Smith: Quantum confinement effects of semiconducting microcrystallites in glass. *J. Appl. Phys.*, **61** (1987), 5399-5409

36) P. Roussignol, D. Ricard, J. Lukasik and C. Flytzanis: New results on optical phase conjugation in semiconductor-doped glasses. *J. Opt. Soc. Am B*, **4** (1987), 5-13

37) H. Nasu and J. D. Mackenzie: Nonlinear optical properties of glasses and glass- or gel-based composites. *Opt. Eng.*, **26** (1987), 102-106

38) K. C. Rustagi and J. Ducuing: Third-order optical polarizability of conjugated organic molecules. *Opt. Commun.*, **10** (1974), 258-261

39) C. Sauteret, J. P. Hermann, R. Frey, F. Paradere, J. Ducuing, R. H. Baughman and R. R. Chance: Optical nonlinearities in one-dimensional-conjugated polymer crystals. *Phys. Rev. Lett.*, **36** (1976), 956-959

40) G. M. Carter, Y. J. Chen and S. K. Tripathy: Intensity dependent index of refraction in organic materials. *Opt. Eng.*, **24** (1985), 609-612

41) Fujiwara and Nakagawa: Phase-Conjugate Optics Based on Organic Dye Films. *Optics*, **18** (1989), 122-131.

Chapter 8

1) D. A. Rockwell: A review of phase conjugate solid-state lasers. *IEEE J. Quantum Electron.*, **24** (1988), 1124-1140

2) C. R. Giuliano: Applications of optical phase conjugation. *Phys. Today*, **34** (1981), 27-35

3) J. AuYeung, D. Fekete, D. M. Pepper and A. Yariv: A theoretical and experimental investigation of the modes of optical resonators with phase-conjugate mirrors. *IEEE J. Quantum Electron.*, **QE-15** (1979), 1180-1188

4) J. Feinberg and R. W. Hellwarth: Cover photo of Laser Focus, December 1981 issue.

5) T. R. O'Meara: Compensation of laser amplifier trains with nonlinear conjugation techniques. *Opt. Eng.*, **21** (1982), 243-251

6) M. D. Levenson, K. M. Johnson, V. C. Hanchett and K. Chiang: Projection photolithography by wave-front conjugation. *J. Opt. Soc. Am.*, **71** (1981), 737-743

7) S. Kawakami and J. Nishizawa: An optical wavequide with the optimum distribution of the refractive index with reference to waveform distortion. *IEEE Trans. Microwave Theory Tech.*, **MTT-16** (1968), 814-818

8) A. Yariv: On transmission and recovery of three-dimensional image information in optical waveguides. *J. Opt. Soc. Am.*, **66** (1976), 301-306

9) A. Yariv: Three-dimensional pictorial transmission in optical fibers. *Appl. Phys. Lett.*, **28** (1976), 88-89

10) G. J. Dunning and R. C. Lind: Demonstration of image transmission through fibers by optical phase conjugation. *Opt. Lett.*, **7** (1982), 558-560

11) F. A. Hopf: Interferometry using conjugate-wave generation. *J. Opt. Soc. Am.*, **70** (1980), 1320-1323

12) I. Bar-Joseph, A. Hardy, Y. Katzir and Y. Silberberg: Low-power phase-conjugate interferometry. *Opt. Lett.*, **6** (1981), 414-416

13) T. Sato, T. Suzuki, P. J. Bryanston-Cross, O. Ikeda and T. Hatsuzawa: Coherent optical image delay device using a BSO phase-conjugate mirror and its applications. *Appl. Opt.*, **22** (1983), 815-818

14) P. Yeh, I. McMichael and M. Khoshnevisan: Phase-conjugate fiber-optic gyro. *Appl. Opt.*, **25** (1986), 1029-1030

15) D. M. Pepper, J. AuYeung, D. Fekete and A. Yariv: Spatial convolution and correlation of optical fields via degenerate four-wave mixing. *Opt. Lett.*, **3** (1978), 7-9

16) B. H. Soffer, G. J. Dunning, Y. Owechko and E. Marom: Associative holographic memory with feedback using phase-conjugate mirrors. *Opt. Lett.*, **11** (1986), 118-120

17) A. Yariv and S. Kwong: Associative memories based on message-bearing optical modes in phase-conjugate resonators. *Opt. Lett.*, **11** (1986), 186-188

18) Y. Owechko: Nonlinear holographic associative memories. *IEEE J. Quantum Electron.*, **25** (1989), 619-634

19) H. G. Winful and J. H. Marburger: Hysteresis and bistability in degenerate four-wave mixing. *Appl. Phys. Lett.*, **36** (1980), 613-614

20) D. M. Pepper and R. L. Abrams: Narrow optical bandpass filter via nearly degenerate four-wave mixing. *Opt. Lett.*, **3** (1978), 212-214

21) J. Nilsen, N. S. Gluck and A. Yariv: Narrow-band optical filter through phase conjugation by nondegenerate four-wave mixing in sodium vapor. *Opt. Lett.*, **6** (1981), 380-382

22) R. L. Abrams and R. C. Lind: Degenerate four-wave mixing in absorbing media. *Opt. Lett.*, **2** (1978), 94-96: Errata, **3** (1978), 205

23) G. P. Agrawal: Highly nondegenerate four-wave mixing in semiconductor lasers due to spectral hole burning. *Appl. Phys. Lett.*, **51** (1987), 302-304

24) H. Nakajima and R. Frey: Intracavity nearly degenerate four-wave mixing in a (GaAl) As semiconductor laser. *Appl. Phys. Lett.*, **47** (1985), 769-771

25) D. G. Steel and R. C. Lind: Multiresonant behavior in nearly degenerate four-wave mixing: the ac Stark effect. *Opt. Lett.*, **6** (1981), 587-589

26) D. J. Harter and R. W. Boyd: Nearly degenerate four-wave mixing enhanced by the ac Stark effect. *IEEE. J. Quantum Electron.*, **QE-16** (1980), 1126-1131

27) A. Yariv, D. Fekete and D. M. Pepper: Compensation for channel dispersion by nonlinear optical phase conjugation. *Opt. Lett.*, **4** (1979) 52-54

28) T. Yajima and Y. Taira: Spatial optical parametric coupling of picosecond light pulses and transverse relaxation effect in resonant media. *J. Phys. Soc. Jpn*, **47** (1979), 1620-1626

29) J. G. Fujimoto and E. P. Ippen: Transient four-wave mixing and optical pulse compression in the femtosecond regime. *Opt. Lett.*, **8** (1983), 446-448

30) D. T. Hon: Pulse compression by stimulated Brillouin scattering. *Opt. Lett.*, **5** (1980), 516-518

31) R. J. Glauber: Coherent and incoherent states of the radiation field. *Phys. Rev.*, **131** (1963), 2766-2788

32) D. F. Walls: Squeezed states of light. *Nature*, **306** (1983), 141-146

33) C. M. Caves and B. L. Schumaker: New formalism for two-photon quantum optics. I. Quadrature phases and squeezed states. *Phys. Rev. A.*, **31** (1985), 3068-3092

34) H. P. Yuen and J. H. Shapiro: Generation and detection of two-photon coherent states in degenerate four-wave mixing. *Opt. Lett.*, **4** (1979), 334-336

35) R. E. Slusher, L. W. Hollberg, B. Yurke, J. C. Mertz and J. F. Valley: Squeezed states in optical cavities: A spontaneous-emission-noise limit. *Phys. Rev. A*, **31** (1985), 3512-3515

36) M. D. Levenson, R. M. Shelby, A. Aspet, M. Reid and D. F. Walls: Generation and detection of squeezed states of light by nondegenerate four-wave mixing in an optical fiber. *Phys. Rev. A*, **32** (1985), 1550-1562

37) M. W. Maeda, P. Kumar and J. H. Shapiro: Observation of squeezed noise produced by forward four-wave mixing in sodium vapor. *Opt. Lett.*, **12** (1987), 161-163

Appendix

1) R. W. Minck, R. W. Terhune and C. C. Wang: Nonlinear optics. *Appl. Opt.*, **5** (1966), 1595-1612

2) Y. R. Shen: "The principles of nonlinear optics," John Wiley & Sons, New York, 1984, p. 28

3) R. W. Hellwarth: Third-order optical susceptibilities of liquids and solids. *Prog. Quantum Electron.*, **5** (1977), 1-68

4) J. F. Nye: "Physical properties of crystal," Clarendon, Oxford, 1957, 295-301

5) P. M. Morse and H. Feshbach: "Methods of theoretical physics, Part 1" McGraw-Hill, New York, 1953, Ch. 3

6) "Fundamentals of Modern Physics 3," Quantum Mechanics I, Iwanami Booksellers, 1978, p. 593.

Index